NAVIGATING A

A Guide for Women and Mi...

NAVIGATING ACADEMIA

A Guide for Women and Minority STEM Faculty

Edited by

PAULINE MOSLEY

Full Professor of Information Technology, Seidenberg School of CSIS,
Pace University, Pleasantville, NY

S. KEITH HARGROVE

Dean of the College of Engineering, Tennessee State University, Nashville, TN

AMSTERDAM • BOSTON • HEIDELBERG • LONDON
NEW YORK • OXFORD • PARIS • SAN DIEGO
SAN FRANCISCO • SINGAPORE • SYDNEY • TOKYO

Academic Press is an imprint of Elsevier

Academic Press is an imprint of Elsevier
32 Jamestown Road, London NW1 7BY, UK
525 B Street, Suite 1800, San Diego, CA 92101-4495, USA
225 Wyman Street, Waltham, MA 02451, USA
The Boulevard, Langford Lane, Kidlington, Oxford OX5 1GB, UK

ISBN: 978-0-12-801984-9

British Library Cataloguing-in-Publication Data
A catalogue record for this book is available from the British Library

Library of Congress Cataloging-in-Publication Data
A catalog record for this book is available from the Library of Congress

For Information on all Academic Press publications
visit our website at http://store.elsevier.com/

Typeset by MPS Limited, Chennai, India
www.adi-mps.com

Dedication

This book is dedicated to our families. Without their support, this work would not be possible. We thank you for your understanding, in spite of all the time it took us away from you.

Acknowledgments

I would like to express my gratitude to the many people who saw us through this book. Special thanks to Mary Preap for helping us in the process of selection and editing.

S. Keith Hargrove

My determination, strong optimism, and spiritual strength I owe to the wonderful women in my life: my darling grandmother, my dear mother, and all the wonderful church mothers who showed me how to look to God to overcome any obstacle or barrier life brings.

Above all I want to thank my husband, Paul, for financially supporting this dream to develop a book which encourages minorities and women to never give up - until success is achieved!

Pauline Mosley

Contents

Section A
ISSUES

Section B
INSTITUTIONAL AND SOCIETAL CULTURES

Section C
CAREER PATHWAYS

Section D
TRANSITIONING

Section E
MENTORING

Section F
PROFESSOR SUPPORT NETWORKS

About the Editors

DR. PAULINE MOSLEY

Dr. Pauline Mosley holds a Bachelor of Science in maths and a Bachelor of Science in computer science from Mercy College, a Master of Science in information systems, and a Doctorate on professional studies from Pace University, New York City, NY. She embarked upon a teaching career in 1986, working as a top corporate trainer for Personal Computer Learning Centers of America, Inc. where she trained Fortune 500 executives and personnel in a myriad of software applications. She developed computer training manuals for Texaco, Pepsi, The Port Authority, and McCraw-Hill and was influential in establishing PC and mainframe user-support help desks for Dannon, NYNEX, and Brooklyn Union Gas.

Prior to joining Pace in 2000, she was a tenured City University of New York faculty member for 10 years and an adjunct professor at the following colleges: Westchester Community College, Iona, The College of New Rochelle, and Mercy College. She is the recipient of Who's Who Among America's Teachers and Women in Technology Award from the Young Women's Christian Association (YWCA). She is a full professor of information technology in the School of Computer Science and Information Systems at Pace University in Pleasantville. She teaches LEGO robotics, web design, internet and network security, database management, systems analysis, and design and various service-learning courses. Her research interests include cognitive models for STEM learning robotics and web development. She is the recipient of several National Science Foundation grants that have enabled her to pursue her research interests which involve exploring pedagogical STEM methodologies that increase minority and young women participation and designing secure systems. She is a member of the Institute of Electrical and Electronic Engineers, Inc. (IEEE) and frequently serves on the program committee of National Conferences on Information Technology. She is an active board member for the YWCA in White Plains, and the Youth Mission of Life, as well as an advisor to the White Plains GIRLS Academy. Journals in which her research has appeared include *The Journal of Computing Sciences in Colleges, Across The Disciplines,* and *The Academic Exchange Quarterly.*

DR. S. KEITH HARGROVE

Dr. S. Keith Hargrove serves as Dean of the College of Engineering at Tennessee State University. He received his BS in mechanical engineering from Tennessee State University, Nashville, TN; MS from the Missouri University of Science & Technology in Rolla, MO, as a GEM Fellow; and PhD from the University of Iowa as a CIC Fellow. He previously served as chairperson of the Department of Industrial, Manufacturing & Information Engineering in the Clarence Mitchell, Jr. School of Engineering at Morgan State University in Baltimore, MD, and as assistant to the dean and associate professor of Mechanical Engineering in the College of Engineering, Architecture & Physical Sciences at Tuskegee University, Tuskegee, AL. He has worked for General Electric, Battelle Pacific Northwest Laboratory, NIST, Oak Ridge Laboratory, General Motors, and as a research professor at the University of Michigan. He has received research funding from the National Science Foundation and conducted research projects with Sikorsky Aircraft, Boeing, NASA, US Navy, and the US Army in systems engineering, design, virtual and augmented reality, advanced manufacturing, and minority engineering education. He is the director of the TIGER Research Institute at Tennessee State University, a group of laboratories funded by external grants and contracts. He has received several awards for teaching, research, mentoring, and is an associate member of the Society of Manufacturing Engineers, Institute of Industrial Engineers, ASEE, Tennessee Academy of Science, and the Tennessee Society of Professional Engineers. A strong believer in K-12 STEM education, he is a founding board member of STEM Preparatory Academy, a local charter school in Nashville, and is active with curriculum programs at Union Elementary STEM School, Madison Creek Elementary Science Program, and Stratford STEM Magnet High School in Nashville, TN. He is also a strong advocate for mentoring tenure-track and minority faculty throughout their academic careers.

About the Contributors

MARY OLING-SISAY, EdD

Mary Oling-Sisay is the vice president for Student Affairs and dean of Students at St. Norbert College (SNC) in De Pere, WI. She provides vision, strategic direction, and management oversight for a comprehensive student affairs program. She is a 2009–10 Fellow of the American Council on Education.

At SNC, she provides leadership for student-centered programming opportunities designed to enhance retention. Prior to joining SNC in 2007, she held several administrative positions at the City College of the City University of New York, and at California State University, Chico, CA. She has provided leadership in developing initiatives that enhance equity, inclusiveness, and diversity. She has experience in facilitating collaboration between academic affairs and student affairs to attain a positive learning environment for students. She has presented on and written numerous articles on equity, inclusiveness, and diversity; student affairs; and international relations.

She has also had teaching experience in international relations, educational leadership, and women and gender studies. She is affiliated with many higher education organizations and has presented, written, and consulted widely on a broad range of student affairs issues, including student judicial affairs, student culture and diversity, emergency preparedness, student leadership, and campus programming. She has served regularly as a National Association of Student Personnel Administration (NASPA) national conference proposal reviewer. She also serves as an accreditation Consultant Evaluator for the North Central Association/Higher Learning Commission.

She is the president of Wisconsin Women in Higher Education Leadership (WWHEL). Her degrees include a doctorate in educational leadership from the University of Southern California, Los Angeles, CA; an MA in public communications from Fordham University, Bronx, NY; and a BA in English literature and linguistics from Makerere University, Kampala, Uganda. She is also a graduate of the Higher Education Resources (HERS) Institute for Women in Higher Education at Bryn Mawr College, Bryn Mawr, PA.

LENORA ARMSTRONG, BS, MEd

Lenora Armstrong is an assistant professor at Norfolk State University, Norfolk, VA, in the Department of Health, Physical Education and Exercise Science for the past 11 years. She obtained her BS from Hampton University, Hampton, VA; MEd from Temple University, Philadelphia, PA; CAS from Illinois State University, Normal, IL. She is currently pursuing doctorate degree in Higher Education Administration at George Washington University, Washington, DC. She has 25 years of teaching experience (3 years—Illinois State University; 10 years—Hampton University; 12 years—Norfolk State University). She is a member of AAHPERD and VAHPERD and have presented at both conferences on my dissertation topic "Career Pathways of Chief Athletic Administrators." She is working in higher education for 24 years since graduate school teaching at Illinois State University for 3 years, teaching, coaching women's track and field and cheerleading, and the senior women's administrator at Hampton University for 10 years. She is in the NACWAA/HERS Institute 2003 East Class. In her leisure time, she enjoys working out, traveling, the beach, and a variety of entertainment.

KIMBERLY M. COLEMAN, PhD, MPH, CHES

Dr. Kimberly Coleman is a Certified Health Education Specialist (CHES) with approximately 10 years of public health and health education experience, including health ministry management, health care services administration, research project coordination, smoking cessation and weight management-focused health counseling, and community-based participatory research. Her research and practice expertise include examining the relationship between religion, spirituality, and health; development of programs designed to promote health equity and social justice; and HIV/AIDS prevention education. Most recently, she was as an assistant professor of community health at Georgia Southern University from 2006 through 2010, where she was the first joint appointed faculty member between the College of Health and Human Sciences and Jiann-Ping Hsu College of Public Health. In 2008, she was awarded a 2-year, $100,000 contract from the W.K. Kellogg Foundation providing Technical Assistance Coordination for the "New Tools, New Visions 2" project. This project partnered four rural southern Georgia communities and Historically Black Colleges and Universities to develop community-based participatory research infrastructures. She is a coauthor of a chapter on leisure activity and spirituality in *Leisure, Health and Wellness: Making the Connections* (in press). She has also presented original works of research at the annual meetings of the American Public Health Association (APHA) and American Alliance for Health, Physical Education, Recreation, & Dance. She has served as an active member in several public health and health education-focused organizations, including the national health education honorary, Eta Sigma Gamma, the American Public Health Association, the American Association for Health Education, and a past governing board member for the Society for African-American Public Health Issues. She currently holds leadership positions with the American Association for Health Education (AAHE) Membership Committee and the Caucus on Public Health and the Faith Community (APHA).

A native of Southeast Washington, DC, she is an alumna of Spelman College, Atlanta, GA, earning a BA in psychology in 1994. She earned the MPH in health behavior and health education in 2002 and immediately began doctoral studies in health education at Southern Illinois University Carbondale (SIUC), Carbondale, IL. While at SIUC, she worked on a number of health education and health behavior interventions and research projects, including assessing the perceived needs of African-American Christian women during pregnancy and childbirth, developing a smoking cessation intervention for pregnant adolescents, and evaluating the Southern Illinois Healthcare Parish Nursing Program. Her dissertation research focused on examining the behavioral factors that motivate African-American Christians to participate in HIV/AIDS ministries, earning the PhD in 2006. She ultimately dreams of returning to the District of Columbia and creating a nondenominational health ministry clearinghouse that will "impact the mental, physical, social, environmental, and spiritual health of individuals and communities by designing programs that are culturally and spiritually responsive."

KERA Z. WATKINS, PhD

Kera Z. Watkins grew up in the Washington, DC, metropolitan area. The author received her BS in mathematics from Spelman College, Atlanta, GA, in 1996. She then earned her MS in computer science from Clark Atlanta University, Atlanta, GA, in 1999. She finally earned her PhD in computer science from North Carolina State University, Raleigh, NC, in 2006.

She is currently an assistant professor in the Department of Computer Sciences at Georgia Southern University (GSU) in Statesboro, GA. She is also the director of the Software Testing Laboratory at GSU. She conducted software engineering research as a NASA Jenkins Pre-doctoral Fellow. She is currently GSU's College of Information Technology (CIT) Academic Liaison for the Students and Technology in Academia, Research, and Service (STARS) Alliance (sponsored by NSF). She has been active in using software tools to excite K-12 and college students towards computing. She has a broad interest in software engineering, particularly in software testing in complex networking environments. She also has an interest in enhancing the computer science and software engineering educational experience. Some recent publications include the following: "Introducing Fault-Based Combinatorial Testing to Web Services," Proceedings of the 2010 IEEE SoutheastCon, March 2010; *Towards Minimizing Pair Incompatibility to Help Retain Under-represented Groups in Beginning Programming Courses*, ACM Journal of Computing Sciences in Colleges, December 2009; and, a number of other publications.

She has been nominated for GSU's 2010 CIT Dean's Citation for Service. She received the 2009 CIT Dean's Citation for Service and was nominated for GSU's 2009 Excellence in Service Award. While serving as the Academic Liaison for the STARS Alliance, GSU was named the "STAR of the month" in January 2009.

RAQUEL DIAZ-SPRAGUE, PharmD

Originally from Trujillo, Peru, former Fulbright Scholar, **Raquel Diaz-Sprague**, PharmD MS MLHR, is a biochemist with a wide range of intellectual interests. She is known for her commitment to increasing participation and equity for women in science, promoting intercultural understanding and respect for the Latino community, and to eliminating barriers and

disparities. She has been an editor of journal articles in biochemistry with Chemical Abstracts, language consultant with Technical Support Inc., adjunct faculty in the Ohio State University, College of Medicine, and the director of the Women in Science Day program hosted by OSU since 1999. She is the founding director of a free clinic for uninsured Latinos and has taught medical communication with Latinos since 1992, first in the Medical Humanities Program of the College of Medicine and more recently as an adjunct instructor in the School of Allied Medical Professions. She has facilitated medical students' visits to Trujillo, Peru, since 1993 and has convened the Ethics in Science, Technology, and Medicine conference series at OSU since 1991. She is a frequent con- vener or presenter at meetings of the Association for Practical & Professional Ethics and a case writer for the Inter-Collegial Ethics Bowl organized by the Illinois Institute of Technology. For her leadership and service she has been named an Exemplar by the Ohio Academy of Science, honored as a Distinguished Alumna for Outstanding Community Service by Ohio State University, and elected Fellow by the Association for Women in Science. She was chosen as the 2006 Commencement Speaker for Columbus State Community College and has been inducted in the Ohio Women's Hall of Fame for Achievements in Math, Science, and Health Services. She has received over 100 small to medium size grants to fund dozens of conferences and service pro- jects. She is the author or coauthor of numerous articles on women and girls in science, ethical issues, and Latino perspectives on health issues. She translated and edited the book: "Tu Salud y La Ley: Una Guia Para Adolescentes" published by the ACLU of Ohio in 2007. One of her lec- tures on Health Communication is part of the Supercourse on Epidemiology and Global Health hosted by the University of Pittsburg http://www.pitt.edu/~super1/.

DIXIE L. DENNIS, PhD, CHES

Dixie L. Dennis became the associate provost for Grants and Sponsored Programs and dean of the College of Graduate Studies at Austin Peay State University (APSU) in February 2010. Before then and since she obtained her PhD in 2000, she served as director of the Title III Grant, director of Grants and Sponsored Research. Prior to those positions, she was dean of the College of Science and Mathematics, dean of the College of Professional Programs and Social Sciences, and chair of the Department of Health and Human Performance—all at APSU. Before coming to APSU, she served as Research Coordinator of two doctoral programs at the University of Maryland Eastern Shore.

She has given much time in service to the community, university, and her profession. For example, she is the immediate past chair of the National Commission for Health Education Credentialing and just finished service on the Board of Directors for the American Association for Health Education. Throughout this service to her profession, she served as a contributing editor for the *American Journal of Health Education*, reviewed for numerous journals, and was named the recipient of the Richard M. Hawkins award for research at APSU.

She is the author and coauthor of numerous peer-reviewed journal publications, with a research focus of spirituality. She has authored two book chapters and a textbook, *Living, Dying, Grieving*. Since 1998, she has delivered scholarly and professional presentations at national conferences and conventions.

ROSEMARIE TILLMAN, PhD

Rosemarie Tillman finds grounding for her research focus on faculty work life, and how collegiality, communication, mentoring relationships, policy, diversification of American college and university faculty impact a scholarly career through a natural synthesis of her assorted experiences. She received an AAS degree (1992) from Onondaga Community College in Radio/ Television; multiple degrees from Syracuse  University—a BS degree (1993) and an MS degree (1995) in speech communication and an MS degree (1996) and CAS (1996) in instructional design, development and evaluation; and a PhD degree (2006) from the University of Oregon in Educational Policy and Management.

Since the awarding of her doctoral degree, she has increased her participation within professional associations, presenting four professional papers (three to the American Educational Research Association [AERA] and one to the Western States Communication Association [WSCA]) during the last 4 years. At the close of the 2008 academic year, she found conducting research, giving presentations, pursuing publication, guest lecturing, and adjunct teaching without the support of a full-time academic home to be counterproductive. Consequently, she has declined adjunct appointments in order to more rigorously pursue research projects and a tenure-track post.

LAVAR J. CHARLESTON, PhD

LaVar J. Charleston is a PhD candidate in the department of Educational Leadership and Policy Analysis at the University of Wisconsin-Madison. He is also a research associate at the Wisconsin Center for Education Research. He is a Wisconsin Spencer Doctoral Research Fellow and a School of Education Graduate Research Scholar. His research focuses on broadening participation for underrepresented groups in Science, Technology, Engineering, and Mathematics. He has received numerous awards including the Tom W. Shick Scholarship, Henry Arnsbrak Graduate Fellowship, and the Golden Key Education Achievement Scholarship. Additionally, he serves as the associate editor for the *Annuals of the Next Generation* Journal and he is the assistant director for development and marketing at the Center for African American Research and Policy. He serves as an assistant evaluator for the African American Researchers in Computer Science program, a National Science Foundation funded project. He has published on a variety of topics including increasing the participation of African-Americans in the computing sciences, assessing regional STEM participation among underrepresented populations, mentoring African-American graduate students within the computing sciences, and examining higher education access for underserved populations post anti-affirmative action legislation. He has presented his work at several professional conferences including the American Educational Research Association, the Association for the Study of Higher Education, and the Think Tank for African American Progress.

JERLANDO F. L. JACKSON, PhD

Jerlando F. L. Jackson is an associate professor of Higher and Postsecondary Education in Educational Leadership and Policy Analysis at the University of Wisconsin-Madison. He also serves as the Coordinator for the Higher, Postsecondary, and Continuing Education Program. He is a faculty affiliate for both the Wisconsin Center for the Advancement of Postsecondary Education (School of Education)

and the Weinert Center for Entrepreneurship (School of Business). His central research interest has been to explore workforce diversity and workplace discrimination in higher education. In addition, he serves as the executive director for the Center for African American Research and Policy, which is developing and publishing a new generation of research on policy issues confronting African-Americans in both the academy and the society at-large. Frequently sought as a keynote speaker, he is credited with over 75 publications, 100 presentations, and has published the following books—*Ethnic and Racial Administrative Diversity: Understanding Work Life Realities and Experiences in Higher Education* for Jossey-Bass (2009); *Strengthening the African American Educational Pipeline: Informing Research, Policy, and Practice* for SUNY-Albany Press (2007); and *Toward Administrative Reawakening: Creating and Maintaining Safe College Campuses* for Stylus Publishing (2007). In addition, he conducts research on and evaluations of interventions designed to broaden participation for underrepresented groups in STEM fields. Currently, he is the lead evaluator for four National Science Foundation funded projects: (i) African American Researchers in Computing Sciences; (ii) PC2Main; (iii) Software Design and Development Teams Tournament; and (iv) Alliance for the Advancement of African American Researchers in Computing.

JUAN E. GILBERT, PhD

Juan E. Gilbert is a professor and chair of the Human-Centered Computing Division in the School of Computing at Clemson University where he leads the Human-Centered Computing (HCC) Lab. He has research projects in spoken language systems, advanced learning technologies, usability and accessibility, Ethno computing (Culturally Relevant Computing), and databases/data mining. He has published more than 90 articles, given more than 140 talks and obtained more than $9 million dollars in research funding. He was recently named one of the 50 most important African-Americans in Technology. He was also named a Speech Technology Luminary by Speech Technology Magazine and a national role model by Minority Access Inc. He is also a national associate of the National Research Council of the National Academies, an ACM Distinguished Speaker, a member of the IEEE Computer Society Distinguished Visitors Program, and a senior member of the IEEE Computer Society. Recently, he was named a Master of

Innovation by Black Enterprise Magazine, a Modern-Day Technology Leader by the Black Engineer of the Year Award Conference, the Pioneer of the Year by the National Society of Black Engineers and he received the Black Data Processing Association (BDPA) Epsilon Award for Outstanding Technical Contribution. In 2002, he was named one of the nation's top African-American Scholars by Diverse Issues in Higher Education.

List of Contributors

Ryan P. Adserias Wisconsin's Equity and Inclusion Laboratory (Wei LAB), University of Wisconsin-Madison, WI

Lenora Armstrong Department of Health, Physical Education & Exercise Science, Norfolk State University, Norfolk, VA

LaVar J. Charleston Wisconsin's Equity and Inclusion Laboratory (Wei LAB), University of Wisconsin-Madison, WI

Kimberly M. Coleman College of Public Health, Georgia Southern University, Statesboro, GA

Dixie L. Dennis Austin Peay State University, Clarksville, TN

Raquel Diaz-Sprague Technical Support, Inc., Columbus, OH

Juan E. Gilbert Computer & Information Science & Engineering Department, University of Florida, FL

S. Keith Hargrove College of Engineering, Tennessee State University, Nashville, TN

Jerlando F. L. Jackson Department of Educational Leadership and Policy Analysis & Wisconsin's Equity and Inclusion Laboratory (Wei LAB), University of Wisconsin-Madison, WI

Pauline Mosley Seidenberg School of CSIS, Pace University, Pleasantville, NY

Mary Oling-Sisay Alliant International University, San Diego, CA

Rosemarie Tillman Linn-Benton Community College, OR and Lane Community College, OR

Kera Z. Watkins Department of Computer Science, Georgia Southern University, Statesboro, GA

Preface

The future of higher education is being challenged by several internal and external factors that will determine its true educational value, impact on quality of life, and digital tools for learning. These elements will be transformed by the issues of accountability, fiscal sustainability, learning methodologies, student quality, and leadership. The authors believe that leadership is the most critical as it dictates the decision making of the institution to transform itself in meeting the needs of customers (students), and the autonomy of faculty to deliver knowledge, and grow intellectually and professionally in the ivory tower of the academy.

The potential reader of this book may have been captured by the title. Or possibly, a quest to seek information and advice on how to succeed in higher education as a faculty member, staff personnel, or administrator. In fact, one may find a number of well-written books that provide knowledge on how institutions of higher learning operate; how to effectively run them; and how to navigate through the processes of tenure, politics, and academic cultures. They all have their purpose, and so is the case for this collection of writings. However, juxtapose to the experience of the majority of academic professionals in academia, the subpopulation of minority/underrepresented groups may tend to have a somewhat different career livelihood than the majority group. This text focuses on that particular group and their experiences.

The authors can relate to aforementioned issues and were primarily motivated to share reflections of their experiences and others of the minority academic population. We believe that through the sharing of these experiences may help others survive, rebound, plan, or just simply advance their academic careers with some collegial advice on having a career in the academy as we know it. Thus, an honest attempt has been made to share the reflections of several minority colleagues from academia. This text begins with an overview of issues and a generic profile of employees of higher education, providing a personal reflection on the cultural environment of higher education. The reader then gets a narrative of the career pathways of faculty and administrators, followed by reflections on survival tactics and best practices for career progression. In addition, there is a companion website, http://booksite.elsevier.com/9780128019849, for this book with further resources for readers.

As in most organizations, the career success is typically a function of mentoring, networks, performance, and qualifications. Several of the contributors also share this perspective.

Though we recognize that higher education is faced with numerous challenges, we believe that the reader's potential to navigate through the academic maze for leadership advancement will play a critical role in maintaining the US position as an intellectual icon of formal education and training. We also believe a diversity of perspectives is essential as the population slowly migrates from a minority to majority state. Beyond that, to transform today's institution to enhance the quality of learning will require "all hands on deck" to compete globally and the same opportunity for all to advance to leadership based on credentials and scholarly productivity. In your pursuit of academic leadership, we hope that these articles of reflections will help the reader avoid some obstacles, and help prepare and plan to achieve greater heights of leadership in academia or elsewhere.

Pauline Mosley
S. Keith Hargrove

ISSUES

The Issues and Demographic Data

Mary Oling-Sisay

Alliant International University, San Diego, CA

A persistent problem in American higher education is the question of who has access. This matter is especially troublesome when one looks at the composition of senior administrators in most colleges and universities. While some women who occupy leadership positions have broken and are still breaking the glass ceiling, much work needs to be done to increase the number of women in the ranks of senior leadership in higher education (American Council on Education [ACE], 2007c, 2008b; Appadurai, 2009; Corrigan, 2002). This need is evidenced by numerous newspaper articles that announce "she is the first woman" or "the only woman" to hold this position (Turner, 2000, 2007). Women and people of color are still the exception when it comes to presidencies (ACE, 2008b, 2013).

Despite some progress and a plethora of research and empirical studies on gender and leadership in higher education, these studies tend to focus on the recruitment and retention of students and faculty of color and not on administrators (Arredondo, 1996; Barak et al., 1998; Barak, 1999, 2000). Considerable progress has been made on the scholarship about women in higher education, but women of color in senior leadership continue to be underrepresented and various reasons account for this state of affairs (ACE, 2008b, 2013; Corrigan, 2002). This chapter offers a descriptive analysis of the intersection of gender, race/ethnicity, and leadership in higher education from the perspective of a female administrator of color.

Many colleges and universities in the United States started out as institutions by men and for men; thus many colleges and universities emerged as single-sex institutions. Women only formally began to enter higher education in the 1830s. Even then, men and not women were viewed as suitable guardians of academic standards or as apt candidates for presidents of institutions. A few changes occurred in the twentieth century enabling some women to gain access to the position of president or chancellor in selected institutions. Women initially had access to leadership positions in some women's colleges in female-dominated fields such as home economics, nursing, and later dean of women. For persons of color, the experience was and is still somewhat more complex (Barbour, 2008; Touchton, 2008).

For women of color this epoch was a "double-edged sword." In a way, it enabled the entry of a few women of color to the presidency, but at the same time, it also accounts for the underrepresentation of women of color in senior higher education administration (ACE, 2007c, 2008b, 2013; Berryman-Fink et al., 2003; Rolle et al., 2000; Turner and Myers, 2000; Watson, 2001). Figure 1.1 highlights the gender stratification.

This state of affairs has serious implications for academe for it is difficult to see how academe could be immune from the impact of changing demographics (Betances et al., 2006; Cross, 2000; Dass and Parker, 1999). Senior leaders exercise considerable power to shape the discourse around institutional vision (Schein, 1987, 1992, 2004; Bolman and Deal, 2003, 2008). Thus, institutions reflect the interests, needs, and values of its senior leadership. Leaders may be from the dominant group and they may not be attuned to or willing to place matters of inclusivity at the forefront of the institution's agenda (Trepagnier, 2006). These leaders unintentionally or strategically may not communicate values that align with inclusivity and diversity (Dass and Parker, 1999; Ely and Thomas, 2001; Jackson and Jones, 2001; Johnson, 2006). They only serve to continue the marginalization and devaluing of the interests of men and women of color and of White women (Cox, 2001; Johnson, 2006; June, 2007; Marcus, 2000). ACE (2008b) compiled data on the

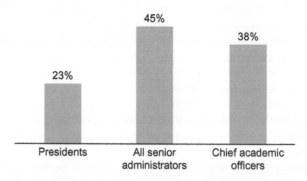

FIGURE 1.1 Percentage of presidents and senior administrators who are female. *Source: From American Council on Education, 2007b. Copyright 2007 by ACE. Reprinted with permission. ACE (Forthcoming). On the Pathway to Presidency: Characteristics of Higher Education's Senior Leadership.*

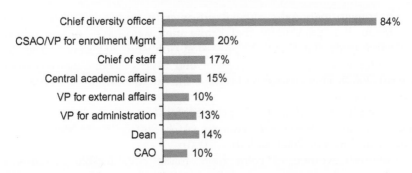

FIGURE 1.2 Percentage of senior administrators who are people of color. *Source: From American Council on Education, 2008a. Copyright 2008 by ACE. Reprinted with permission.*

distribution of people in senior leadership roles, and Figure 1.2 shows the distribution of people of color in leadership roles.

A prevailing perception is the notion that diversity and social justice are issues that are relevant only to men and women of color and to White women, and not to White men. This perception is evidenced by the dearth of White men active in diversity-related work in higher education, and the mild surprise felt by most people when they encounter a White male involved in diversity work (Turner, 2000). The default position at some institutions, especially at predominantly White institutions, is the prevalence of women and men of color and White women being called upon to serve on diversity committees, mentor students of color or female students, and fulfill other functions that serve the institution's diversity goals (Holmes, 2003; Turner and Myers, 2000).

Some senior leaders in higher education view acting in these roles as an expedient way to advance diversity in senior leadership ranks (Gregory, 2003; Holmes, 2003; Jackson and Jones, 2001). It is unfair and unjust to expect people of color or women to be naturally interested in assuming such leadership roles. This expectation places an undue burden on people from groups that are already operating from a marginalized context in the academy (Smith, 2000; Trepagnier, 2006). Diversity and inclusivity should be everyone's concern. It should be about challenging institutional beliefs, values, policies, and practices to enhance inclusiveness (Davis, 2007; Hollis, 2007; Mor Barak and Cherin, 1998).

There is a phenomenon where people of color and women are assigned or when they themselves choose to engage in diversity work because no one else will. In addition, some pursue diversity positions because it is the only way to gain access to a career in the academy (Creed and Scully, 2000; Farmer, 1993; Jackson and Jones, 2001; Kendall, 2006; Sullivan, 2006). This phenomenon is what Padilla (1994) refers to as a "cultural tax."

For transformational change about diversity and inclusion to occur in academia, it is paramount that White men who have well-developed, reflective understandings of oppression (Johnson, 2000) take on leadership roles in diversity work (Turner and Myers, 2000; Villalpando and Bernal, 2002). This change will reduce the unfair burden on women and people of color.

As depicted in Figure 1.2, people of color still tend to be heavily located in diversity-related positions (ACE, 2007c, 2008b; Marcus, 2000; Turner and Myers, 2000; Kolodny, 2000).

Although women of color have made considerable progress in acquiring and serving successfully in senior leadership positions, the struggle continues. Thus, to ensure a pipeline, preparation and inspiration must start at an earlier stage (Cameron and Quinn, 2006; Corrigan, 2002). For women of color, the double-edged sword is very much prevalent especially at predominantly White institutions (Jackson and Jones, 2001; Johnson, 2006). For instance, the number of African American women in the ranks of the faculty and administration are still low compared to their counterparts (ACE, 2007c, 2008b; NCES, 1993, 1994). Despite numerous research analyses and some progress in access to leadership positions, women of color still comprise very low numbers in higher education leadership at the senior or executive level (ACE, 2007c, 2008b, 2013) (Figure 1.3).

Women of color have a difficult time not only because of their gender, but also because of their race and ethnicity (Turner, 2000). Some women of color also have the added challenge of coming from low socioeconomic backgrounds. This multilevel subordination of minority women is termed matrix of domination (Collins, 2000; Marable, 2001). Although there has been much progress in theorizing leadership in higher education, much of it is centered around traditional gender

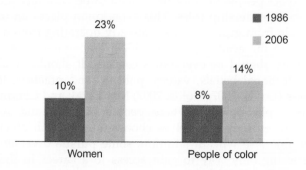

FIGURE 1.3   Women and people of color as a percentage of all presidents: 1986 and 2006. *Source: From American Council on Education, 2007b. Copyright 2007 by ACE. Reprinted with permission.*

models (Gregory, 2003; Kanter, 1993; Schmidt, 2008). These models tend to focus on the highest level of leadership in the academy—the presidency—and they do not fully account for the subtle experiences of women of color in senior-level higher education administration (Holmes, 2003; Watson, 2001).

Black women are often rendered invisible because of their race and gender. However, successful Black women are highly "invisibly visible" at their institutions for a variety of reasons. This phenomenon translates into professional responsibilities that do not add value to their administrative success. These services range from having their faces plastered everywhere to having their decisions, especially those that entail change initiatives, scrutinized intensely.

A related phenomenon is assuming the role of "resident authority" (Turner and Myers, 2000; Villalpando and Bernal, 2002). This phenomenon refers to the expectation placed on female administrators of color to manage all affairs related to people of color (Sullivan and Tuana, 2007). In addition, the "house guest" syndrome can accompany the resident authority phenomenon. This syndrome is where the female administrator of color is expected to be visible, but the actions of those around her render her invisible (Davidson, 1999). This syndrome has a sense of "temporariness" that is accorded to house guests (Holmes, 2003; Turner, 2000; Watson, 2001). According to ACE (2007c, 2008b, 2013), White males still constitute the majority of senior leadership in higher education. Figure 1.4 illustrates recent percentages.

Another phenomenon is the "lone soldier." Often, women of color find that they are the only female of color in senior administration. This lone soldier faces very trying times because she is expected to assimilate and to assume the characteristics of the dominant group (Collins, 2000; Ely and Thomas, 2001; June, 2007; Stubblefield, 2005;

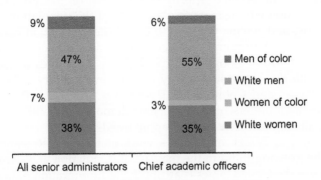

FIGURE 1.4   White men continue to dominate the senior leadership roles. *Source: From American Council on Education, 2007a. Copyright 2007 by ACE. Reprinted with permission.*

Jogulu and Wood, 2006). Expected activities include socialization after hours and water cooler conversations that do not take into consideration the interests of the woman of color. In some instances, there is no recognition of the benefit of cultural competence training for senior leadership. There seems to be an assumption that since everyone is at senior-level leadership, all must have acquired these skills somewhere else.

Ramey (1995) surveyed senior-level African American women from 129 Californian institutions of higher education. The study revealed that racism, sexism, family issues, and perception of incompetence were the barriers they faced in the career paths to their positions.

Lindsay (1999) examined the perceptions of climate at American colleges and universities for four African American women university executives. The participants were three presidents and one provost located at three moderately sized public institutions and one private institution. The study reviewed the institutions' strategic plans, mission statements, catalogues, and university home pages. Again, racism, sexism, and socioeconomic status emerged as key themes.

Rolle et al. (2000) explored the individual experiences of African American administrators at predominantly White institutions in the southeastern United States. The institutions in the study comprised community colleges, private four-year institutions, and public state universities with 10% or less students of color. Themes that emerged included the administrative experience is structured by race; the importance of self-assurance and effective communication skills; and understanding the politics of higher education administration.

Against this backdrop are the shifting demographics of the nation's population and changes in the educational demographics. According to the US Bureau of Census (2004), by 2050, people of color will comprise 49.9% of the nation's population. The Bureau projects that Hispanics and Latinos will become approximately 25% of the population, the Black population will increase to about 15% of the population compared to 13% now, and Asians will rise to 8% of the population compared to the current 4%. Whites are projected to become a numerical minority at 47% by 2050. This demographic trend poses challenges as institutions grapple with the impending reality and the inevitability of being transformed (Passel and Cohn, 2008). Passel and Cohn in a Pew Research Center research underscored the seismic demographic shifting situation. Table 1.1 illustrates recent demographic trends.

According to Passel and Cohn's work at the Pew Research Center, part of the growing trend in the increasing population and demographic dynamics of the United States is attributed to immigration patterns. They further project that if current trends continue, then the population will rise to 438 million in 2050, from 296 million in 2005. Immigrants arriving

TABLE 1.1   US Population, Actual and Projected Population for Years 2008 and 2050

US Census Bureau Data	2008 in Millions	% Population*	2050 in Millions	% Population*	% Increase 2008–2050
Hispanic	46.7	15%	132.8	30%	184.3%
Asian	15.5	5.1%	40.6	9.2%	161.9%
Native Hawaiian and other Pacific Islander	1.1	Not reported	2.6	Not reported	136.3%
American Indians and Alaskan Natives	4.9	1.6%	8.6%	2%	75.5%
Black non-Hispanic	41.1	14%	65.7	15%	59.8%
White non-Hispanic	199.8	66%	203.3	46%	1.8%
Two or more races	5.2	Not reported	16.2	Not reported	211.5%

Population Trends
Source: From Passel and Cohn (2008). Copyright 2008 by Pew Research Center. Reprinted with permission.

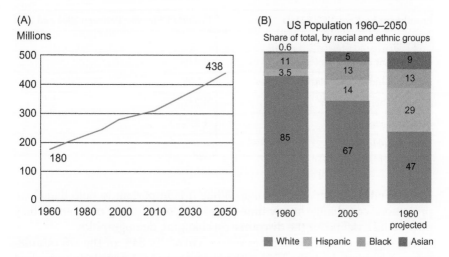

FIGURE 1.5   (A) Actual and projected US population increase and (B) shifts in racial and ethnic groups, 1960–2050. *Source: From Passel and Cohn (2008). Copyright 2008 by Pew Research Center. Reprinted with permission.*

from 2005 to 2050 and their US-born descendants will account for 82% of this increase (Passel and Cohn, 2008) (Figure 1.5).

    Passel and Cohn's projections mirror the US Bureau of the Census projections. The population estimates are based on recent trends and detailed assumptions about births, deaths, and immigration levels—the three key

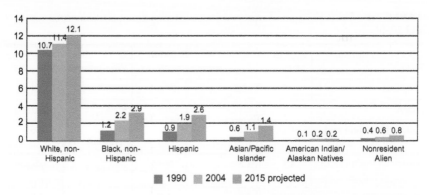

FIGURE 1.6   Projected minority student enrollment increase by race 1990, 2004, and 2015. *Source: From US Department of Education, Spring 2001–Spring 2011; Race/Ethnicity Model, 1980–2010. Copyright 2012 by National Center for Education Statistics. Reprinted with permission.*

TABLE 1.2   Projected Student Enrollment in Higher Education by Race for Years 2004–2015

Enrollment by Race NCES Data	Projected Increase Between 2004 and 2015
Hispanic	+42%
Nonresident Aliens	+34%
American Indian and Alaskan Natives	+30%
Asian or Pacific Islander	+28%
Black non-Hispanic	+27%
White non-Hispanic	+6%

components of population change. While it is important to note that these projections can change at any time depending on a variety of contexts, they offer some baseline for the discourse on changing demographics.

For higher education demographics, currently 34% of the US population, and approximately 35% of the current student population enrolled in higher education, are people of color (Chronicle of Higher Education, 2008). However, there is a disparity when one compares the national population figures with the enrolment figures in higher education and the composition of faculty, administrators, and governing boards. People of color are grossly underrepresented in the following ways:

- 35% of all higher education students (Chronicle of Higher Education, 2008)
- 14% of college and university presidents (ACE, 2007c)

- 19% of executive, managerial, and administrative staff (Chronicle of Higher Education, 2008)
- 22% of full-time faculty members (Chronicle of Higher Education, 2008)
- 25% of part-time faculty members (Chronicle of Higher Education, 2008)
- 22% of governing board members at public colleges and universities (Association for Governing Boards, n.d.)
- 12% of governing board members at independent colleges and universities (Association for Governing Boards, n.d.)

Given the population trends, the enrollment of students of color in higher education is projected to increase tremendously (Chronicle of Higher Education, 2008; Passel and Cohn, 2008) as depicted in Figure 1.6 (Table 1.2).

References

American Council on Education (ACE), 2007a. On the Pathway to the Presidency: 2007, Characteristics of Higher Education's Senior Leadership. ACE, Washington, DC.

American Council on Education (ACE), 2007b. The American College President. ACE, Washington, DC.

American Council on Education (ACE), 2007c. The Graying of the Presidency. ACE, Washington, DC.

American Council on Education (ACE), 2008a. On the Pathway to the Presidency: 2008, Characteristics of Higher Education's Senior Leadership. ACE, Washington, DC.

American Council on Education (ACE), 2008b. The American College President Study: Key Findings and Takeaways. ACE, Washington, DC.

American Council on Education (ACE), 2013. On the Pathway to the Presidency: Characteristics of Higher Education's Senior Leadership. ACE, Washington, DC.

Appadurai, A., 2009. Higher education's coming leadership crisis. Chron. High. Educ. 55 (31).

Arredondo, P., 1996. Successful Diversity Management Initiatives. first ed. Sage Publications, Thousand Oaks, CA.

Barak, M., 1999. Beyond affirmative action: toward a model of diversity and organizational inclusion. Admin. Soc. Work. 23, 47–68.

Barak, M., 2000. The inclusive workplace: an ecosystems approach to diversity management. Soc. Work. 45, 339–353.

Barak, M., Cherin, D., Berkman, S., 1998. Organizational and personal dimensions in diversity climate ethnic and gender differences in employee perceptions. J. Appl. Behav. Sci. 34, 82–104.

Barbour, J., 2008. Organizational culture and institutional discrimination. In: Western Political Science Association, Manchester Hyatt, San Diego, CA.

Berryman-Fink, C., LeMaster, B., Nelson, K., 2003. The women's Leadership program: a case study, *Liberal Education*. Available from: <http://www.aacu-edu.org/liberaleducation/le-wi03/le-wi03Perspective.cfm> (accessed November 20, 2009).

Betances, S., Torres, L., Souder, L., 2006. The Business Case for Diversity. first ed. Souder, Betances & Associates, Inc., Chicago, IL.

Bolman, L., Deal, T., 2003. Reframing Organizations. third ed. Jossey-Bass, San Francisco, CA.

Bolman, L., Deal, T., 2008. Reframing Organizations. fourth ed. Jossey-Bass, San Francisco, CA.

Cameron, K., Quinn, R., 2006. Diagnosing and Changing Organizational Culture. first ed. Jossey-Bass, San Francisco, CA.

Collins, P.H., 2000. Black Feminist Thought: Knowledge Consciousness and the Politics of Empowerment. Unwin Hyman, New York, NY.

Corrigan, M., 2002. The American College President. first ed. American Council on Education, Washington, DC.

Cox, T., 2001. Creating the Multicultural Organization. first ed. Jossey-Bass, San Francisco, CA.

Creed, W., Scully, M., 2000. Songs of ourselves employees' deployment of social identity in workplace encounters. J. Manage. Inq. 9, 391–412.

Cross, E., 2000. Managing Diversity—the Courage to Lead. first ed. Quorum Books, Westport, CT.

Dass, P., Parker, B., 1999. Strategies for managing human resource diversity: from resistance to learning. Acad. Manage. Exec. 13, 68–80.

Davidson, M., 1999. The value of being included: an examination of diversity change initiatives in organizations. Perform. Improv. Q. 12, 164–180.

Davis, P., 2007. Too busy managing to lead? Public Manage. 89, 28–32.

Ely, R., Thomas, D., 2001. Cultural diversity at work: the effects of diversity perspectives on work group processes and outcomes. Adm. Sci. Q. 46, 229–273.

Farmer, S., 1993. Place but not importance: the race for inclusion in academy, Spirit, Space & Survival. first ed. Routledge, New York, NY, pp. 196–217.

Gregory, R., 2003. Women and Workplace Discrimination. first ed. Rutgers University Press, New Brunswick, NJ.

Hollis, R.B., 2007. Leader-as-teacher: a model for executive development success. Organ. Dev. J. 25. (1) 85–89.

Holmes, S., 2003. Black female administrators speak out: narratives on race and gender in higher education. NASAP J. 6, 47–68.

Jackson, J., Jones, L., 2001. A new test for diversity: retaining African American administrators at predominantly white institutions, Retaining African Americans in Higher Education. first ed. Stylus Pub, Sterling, VA.

Jogulu, U., Wood, G., 2006. The role of leadership theory in raising the profile of women in management. Equal Opportun. Int. 25, 236–250.

Johnson, A., 2006. Privilege, Power, and Difference. first ed. McGraw-Hill, Boston, MA.

June, A., 2007. Presidents: same look, different decade. Chron. High. Educ. 5, 33.

Kanter, R.M., 1993. Men and Women of the Corporation. Basic Books, New York, NY.

Kendall, F., 2006. Understanding White Privilege. first ed. Rutledge, New York, NY.

Lindsay, B., 1999. Women chief executives and their approaches towards equity in American universities. Comp. Educ. 35 (2), 187–199.

Marcus, L., 2000. Staff diversity and the leadership challenge. Equity Excell. Educ. 33, 61–67.

Marable, M., 2001. Racism and Sexism. In: Rothenberg, Paula S. (Ed.), Race, Class, and Gender in the United States: An Integrated Study, fifth ed. Worth Publishers, New York, pp. 124–129.

Mor Barak, M., Cherin, D., 1998. A tool to expand organizational understanding of workforce diversity: developing a measure of inclusion–exclusion. Admin. Soc. Work. 22, 47–64.

Padilla, A.M., 1994. Ethnic minority scholars, research, and mentoring: current and future issues. Educ. Res. 23 (4), 24–27.

Passel, J., Cohn, D., 2008. US Population Projections, 2005–2050. first ed. Pew Research Center, Washington, DC.

Race/Ethnicity Model, 1980–2010. Enrollment component; and Enrollment in Degree-Granting Institutions. IPEDS-EF, 96–99.

Ramey, F., 1995. Obstacles faced by African American women administrators in higher education: how they cope. West. J. Black Stud. 19, 113–119.

Rolle, K., Davies, T., Banning, J., 2000. African-American administrators' experiences in predominantly white colleges, universities, and community college. J. Res. Pract. 24, 79–94.

Schein, E., 1987. Organizational Culture and Leadership. first ed. Jossey-Bass, San Francisco, CA.

Schein, E., 1992. Organizational Culture and Leadership. second ed. Jossey-Bass, San Francisco, CA.

Schein, E., 2004. Organizational Culture and Leadership. third ed. Jossey-Bass, San Francisco, CA.

Schmidt, P., 2008. Colleges have blind spots in presidential searches. Chron. High. Educ. 54, A64.

Smith, D., 2000. Women at Work. first ed. Prentice Hall, Upper Saddle River, NJ.

Stubblefield, A., 2005. Ethics Along the Color Line. first ed. Cornell University Press, Ithaca, NY.

Sullivan, S., 2006. Revealing Whiteness. first ed. Indiana University Press, Bloomington, IN.

Sullivan, S., Tuana, N., 2007. Race and Epistemologies of Ignorance. first ed. State University of New York Press, Albany, NY.

Touchton, J., 2008. A Measure of Equity: Women's Progress in Higher Education. first ed. Association of American Colleges and Universities, Washington, DC.

Trepagnier, B., 2006. Silent Racism. first ed. Paradigm Publishers, Boulder, CO.

Turner, C., Myers, S., 2000. Faculty of Color in Academe. first ed. Allyn & Bacon, Boston, MA.

Turner, C.S.V., 2000. New faces, new knowledge: as women and minorities join the faculty, they bring intellectual diversity in pedagogy and in scholarship. Academe. 86 (5), 34–37.

US Department of Education, Spring 2001–Spring 2011. Fall Enrollment Survey National Center for Education Statistics, Integrated Postsecondary Education Data System (IPEDS), IPEDS.

Villalpando, O., Bernal, D., 2002. A critical race theory analysis of barriers that impede the success of faculty of color, The Racial Crisis in American Higher Education. first ed. State University of New York Press, Albany, NY, pp. 243–269.

Watson, L., 2001. In their voices: a glimpse of African-American women administrators in higher education. Natl. Assoc. Stud. Affairs Professionals J. 4, 7–16.

Issues Confronting Athletic Administrators

Lenora Armstrong

Department of Health, Physical Education & Exercise Science,
Norfolk State University, Norfolk, VA

While women hold a healthy number of the total jobs in collegiate athletic administration, they are not well represented in the senior positions. Coakley (2001) argued that jobs for women in coaching and administration are limited because men control most sports programs. Although women's sport programs have increased in number and importance, these teams are still in less powerful positions within the sports hierarchy than those of men. This lesser position connects to the underrepresentation of women at the highest levels of power in sports. Finally, Title IX, which opened the doors for women as participants in sports, seems to have led to closed doors for women in top collegiate sports administration jobs (Abney and Richey, 1992).

Statistics show the low numbers of minority women in Colonial Athletic Association (CAA) roles as 0.6% for Blacks and 0.4% for other minorities in 1995–1996. In 2005–2006, these values change to 1.3% for Blacks and 0% for minorities. In the various categories by divisions, the numbers range from 0% to 2.5% at a Division II Historically Black Colleges and Universities (HBCU) (NCAA, 2007–2008).

Although minority women have made strides in intercollegiate athletics, few have achieved positions in athletic administration in general or in AD positions. Most African American women are concentrated in lower level positions such as secretary, graduate assistant, assistant coach, administrative assistant, assistant athletic director, compliance officers, life skills coordinators, or athletic academic advisor/coordinator (Abney, 2000; Wicker, 2008). Opportunities for minority women have increased in the positions of graduate assistant, academic advisor, senior woman administrator, and intern (NCAA, 2007–2008).

Navigating Academia: A Guide for Women and Minority STEM Faculty.
DOI: http://dx.doi.org/10.1016/B978-0-12-801984-9.00002-X

15

For women to reach equality with men in coaching and athletic administration, women will have to regain lost influence in women's athletics and to enter positions of authority in men's athletics (Feminist Majority Foundation, 1995). According to NCAA (2008–2009), the initial 1989 report on barriers indicated that, in general, women in athletics administration and coaching were content with their careers.

However, some respondents described mixed emotions about their career in intercollegiate athletics. Some described the perception of women as "second-class" citizens. Respondents also indicated that women's involvement with sports was often perceived as an association with lesbian and/or masculine stereotypes. Furthermore, the report highlighted frustrations with politics in athletics and general working conditions. Female administrators and coaches in 1989 attributed the lack of female interest in intercollegiate athletics careers to a perceived interference with marriage and family duties caused by time demands. Other factors highlighted in the report were a lack of initiative for involvement in athletics, stress, lack of advancement and opportunity, and low pay.

The current NCAA (2008–2009) findings indicate that female administrators would still be intercollegiate athletics administrators if they were to start over again. These administrators agree that they encourage current student athletes to consider intercollegiate athletics as a career. While the majority of female administrators indicate satisfaction with their current overall employment, some indicated dissatisfaction with the gender equality within athletics departments and with the equality of race/ethnicity in athletics departments. They also feel there are qualified women who do not apply for intercollegiate athletics administrator positions because of time requirements. In addition, these administrators feel they may leave their careers in athletics administration because of family considerations.

There is little diversity among athletic administrators, and even less diversity at the AD level. Because separate male and female athletic operations merged over the last decade, some areas of diversity have actually decreased. While the overall size of athletic administration has grown, diversity has not advanced faster than this growth. Thus, the problem of practice address is the lack of diversity in athletic administration, especially at the AD level, within colleges and universities. To fully appreciate the AD position, it is important to understand the diversity that exists (or does not exist) currently in these positions.

The highest representation of female athletic directors since the mid-1970s is 21.3% as of 2008. This level represents a significant increase from 18.6% in 2006. In 1972 when Title IX was enacted, females served as athletics directors in over 90% of programs for women. Division III schools have the highest percentage of female athletics directors at

TABLE 2.1 Athletic Directors 1995–1996, 2005–2006, and 2007–2008 (Overall Figures)

Fiscal year	N	White		Black		Other minority		Total		Grand total
		Men	Women	Men	Women	Men	Women	Men	Women	
1995–1996	936	717	140	65	6	7	4	789	150	
2005–2006		X	X	X	X	X	X	872	194	1,066
2007–2008		779	171	75	16	18	7			

Note. The values represent the overall figures of the ADs without the *n* value for years 2005–2006 and 2007–2008. No data reported for overall figures for 2005–2006. Data for 2007–2008 show an increase in the overall figures for all categories.

TABLE 2.2 Athletic Directors 1995–1996, 2005–2006, and 2007–2008 (Overall Percentages)

Fiscal year	N	White		Black		Other minority		Total	
		Men	Women	Men	Women	Men	Women	Men	Women
1995–1996	936	76.4	14.9	6.9	0.6	0.07	0.4	84.0	16.0
2005–2006		72.1	17.9	6.7	1.2	1.7	0.4	80.5	19.5
2007–2008		73.1	16.0	7.0	1.5	1.7	0.7	81.8	18.2

Note. The values represent the overall percentages of the ADs without the *n* value for years 2005–2006 and 2007–2008. A percentage increase in men and a percentage decrease in women with the exception for Black women in 2005–2006. A percentage decrease in men and a percentage increase in women, with a decrease in Black men and an increase in Black women and other minorities.

TABLE 2.3 Athletic Directors 1995–1996, 2005–2006, and 2007–2008—Historically Black Institutions Excluded (Overall Figures)

Fiscal year	N	White		Black		Other minority		Total		Grand total
		Men	Women	Men	Women	Men	Women	Men	Women	
1995–1996	884	709	140	22	2	7	4	738	146	1,011
2005–2006		X	X	X	X	X	X	827	184	
2007–2008		775	171	34	6	18	7	827	184	

Note. The values represent the overall figures of the ADs without the *n* value for years 2005–2006 and 2007–2008. No data given for overall figures for 2005–2006. Data for 2007–2008 show an increase in the overall figures for all categories.

TABLE 2.4 Athletic Directors—Historically Black Institutions Excluded (Overall Percentages) for 1995–1996, 2005–2006, and 2007–2008

Fiscal year	N	White		Black		Other minority		Total	
		Men	Women	Men	Women	Men	Women	Men	Women
1995–1996	884	80.2	15.8	2.5	0.2	0.8	0.5	83.5	16.5
2005–2006		75.8	18.4	3.2	0.3	1.8	0.4	80.9	19.1
2007–2008		82.5	0.7	9.2	1.3	0.0	0.0	91.7	8.3

Note. The values represent the overall percentages of ADs without the *n* value for years 2005–2006 and 2007–2008. A percentage decrease in men and a percentage increase in women with the exception for minority women in 2005–2006. A percentage increase in men and a percentage decrease in women, with an increase in Black men and women and a decrease in other minorities in 2007–2008.

TABLE 2.5   Athletic Directors 1995–1996 and 2007–2008 Figures

Fiscal year	N	White		Black		Other minority		Total		Grand total
		Men	Women	Men	Women	Men	Women	Men	Women	
1995–1996										
Division I	287	236	8	26	3	3	1	265	22	
Division II	265	206	28	26	2	2	1	324	31	
Division III	385	273	94	13	1	2	2	288	97	
2007–2008										
Division I		260	19	37	6	7	2	304	27	331
Division II		208	37	27	8	9	2	244	47	291
Division III		311	115	11	2	2	3	324	120	444

Note. The values represent the figures and the number of the ADs in the divisions without the n value for year 2007–2008 (no data for 2005–2006). The data shows higher numbers and figures for white men in all three divisions with variations of scores of the other categories throughout the divisions.

TABLE 2.6 Athletic Directors 1995–1996, 2005–2006, and 2007–2008 Percentages

Fiscal year	N	White		Black		Other minority		Total	
		Men	Women	Men	Women	Men	Women	Men	Women
1995–1996									
Division I	287	82.2	6.3	9.1	1.0	1.0	0.3	92.3	7.7
Division II	265	77.7	10.6	9.8	0.8	0.8	0.4	88.3	11.7
Division III	385	70.9	24.4	3.4	0.3	0.5	0.5	74.8	25.2
2005–2006									
Division I		82.5	0.7	9.2	1.3	0.0	0.0	91.7	8.3
Division II		67.6	15.7	10.3	2.5	2.9	0.1	80.9	19.1
Division III		72.1	17.9	6.7	1.2	1.7	0.4	80.5	19.5
2007–2008									
Division I		78.5	5.7	11.2	1.8	2.1	0.6	91.8	8.2
Division II		71.5	12.7	9.3	2.7	3.1	0.7	83.8	16.2
Division III		70.0	25.9	2.5	0.5	0.5	0.7	73.0	27.0

Note. The values represent the percentages and the number of the ADs in the divisions without the *n* value for years 2005–2006 and 2007–2008. The data shows higher numbers and percentages for white men in all three divisions with variations of scores of the other categories throughout the divisions in 2005–2006. With the exception of white men, all other categories had an increase in 2007–2008.

TABLE 2.7 Athletic Directors—Historically Black Institutions Excluded 1995–1996 and 2007–2008

Fiscal year	N	White		Black		Other minority		Total		Grand total
		Men	Women	Men	Women	Men	Women	Men	Women	
1995–1996										
Division I	264	232	18	9	1	3	1	244	20	
Division II	240	203	28	6	0	2	1	211	29	
Division III	380	274	94	7	1	2	2	283	97	
2007–2008										
Division I		258	19	19	3	7	2	284	24	308
Division II		206	37	8	2	9	2	223	41	264
Division III		311	115	7	1	2	3	320	119	439

Note. The values represent the percentages of the ADs at HBCUs without the *n* value for year 2007–2008 (no data for 2005–2006). Division I has the highest number and percentage. Division III has the highest scores for women in all women's categories.

TABLE 2.8 Athletic Directors—Historically Black Institutions Excluded 1995–1996, 2005–2006, and 2007–2008 Percentages

Fiscal year	N	White		Black		Other minority		Total	
		Men	Women	Men	Women	Men	Women	Men	Women
1995–1996									
Division I	264	87.9	6.8	3.4	0.4	1.1	0.4	92.4	7.6
Division II	240	84.6	11.7	2.5	0.0	0.8	0.4	87.9	12.1
Division III	380	72.1	24.7	1.8	0.3	0.5	0.5	74.5	25.5
Overall	884	80.2	15.8	2.5	0.2	0.8	0.5	83.5	16.5
2005–2006									
Division I		85.8	7.3	5.0	0.5	1.4	0.0	92.9	7.8
Division II		74.7	17.6	3.3	0.5	3.3	0.5	81.3	18.7
Division III		69.5	26.6	1.9	0.0	1.3	0.6	72.7	27.3
Overall		75.8	18.4	3.2	0.3	1.8	0.4	80.9	19.1
2007–2008									
Division I		83.8	6.2	6.2	1.0	2.3	0.6	92.2	7.8
Division II		78.0	14.0	3.0	0.8	3.4	0.8	84.5	15.5
Division III		70.8	26.2	1.6	0.2	0.5	0.7	72.9	27.1
Overall		76.7	16.9	3.4	0.6	1.8	0.7	81.8	18.2

Note. The values represent the percentages of the ADs at HBCUs without the *n* value for years 2005–2006 and 2007–2008. Division II and III have the highest numbers and percentages with a decrease in Division I. Division III has the highest scores for women in all women's categories.

TABLE 2.9   Athletic Directors—Historically Black Institutions Excluded—Breakdown of Other Minority Category, 2005–2006

Fiscal year	Asian		Hispanic		Native American		Other minority	
	Men	Women	Men	Women	Men	Women	Men	Women
2005–2006								
Division I	0.0	0.0	0.9	0.0	0.0	0.0	0.5	0.0
Division II	0.0	0.5	2.7	0.0	0.5	0.0	0.0	0.0
Division III	0.3	0.3	0.3	0.0	0.6	0.3	0.0	0.0
Overall	0.1	0.3	1.1	0.0	0.4	0.1	0.1	0.0

Note. The data shows the breakdown of percentages of Asian, Hispanic, Native American, and Other minority ADs from each division for 2005–2006. No data reported for 1995–1996.

TABLE 2.10 Athletic Directors Breakdown of Other Minority Categories 2007–2008 (Overall Figures)

Fiscal year	American Indian/ Alaskan Native		Asian		Hispanic/Latino		Native Hawaiian/ Pacific Islander		Other minority		Two or more races		Grand total
	M	W	M	W	M	W	M	W	M	W	M	W	
2007–2008	3	2	1	4	13	1	0	0	0	0	1	0	1,066

Note. Breakdown of American Indian/Alaskan Native, Asian, Hispanic/Latino, Native Hawaiian/Pacific Islander, Other Minority, and Two or More Races for 2007–2008. These figures are more categorized than the 2005–2006 data. No data report for 1995–1996.

TABLE 2.11 Athletic Directors Breakdown of Other Minority Category 2007–2008 (Overall Percentages)

Fiscal year	American Indian/ Alaskan Native		Asian		Hispanic/Latino		Native Hawaiian/ Pacific Islander		Other minority		Two or more races		Grand total
	M	W	M	W	M	W	M	W	M	W	M	W	
2007–2008	0.3	0.2	0.1	0.4	1.2	0.1	0.0	0.0	0.0	0.0	0.1	0.0	

Note. Breakdown of American Indian/Alaskan Native, Asian, Hispanic/Latino, Native Hawaiian/Pacific Islander, Other Minority, and Two or More Races for 2007–2008. These percentages are more categorized than the 2005–2006 data. No data report for 1995–1996.

33.7%. Some schools have no female, at any level, in the athletics administrative structures. The percentage of schools totally lacking a female voice has dropped from 14.5% in 2006 to 11.6 in 2008. The most common administrative structure is composed of three administrators: a male athletics director and one female assistant/associate and one male assistant/associate (Acosta and Carpenter, 2008) (Tables 2.1—2.11).

References

Abney, R., 2000. The glass-ceiling effect and African American women coaches and administrators. In: Brooks, D., Althouse, R. (Eds.), Racism in College Athletics, second ed. Fitness Information Technology, Morgantown, WV, pp. 119—130.

Abney, R., Richey, D., 1992. Opportunities for minority women in sport—the impact of Title IX. J. Phys. Educ. Recreat. Dance. 63, 56—59.

Acosta, V., Carpenter, L., 2008. Women in Intercollegiate Sport. A Longitudinal Study— Twenty-Nine Year Update. first ed. Brooklyn College, Brooklyn, NY, Unpublished manuscript.

Coakley, J., 2001. Sport in Society: Issues and Controversies. seventh ed. McGraw-Hill, New York, NY.

Feminist Majority Foundation and New Media Publishing Inc, 1995. Empowering Women in Sports: The Empowering Women Series No. 4. Available from: <http://www.feminist.org/research/sports4.html.> (accessed May 31, 2009).

NCAA ethnicity and gender demographics of NCAA member institutions' athletic personal. Careers, 2009. National Collegiate Athletic Association.

Wicker, I., 2008. African American Women Athletic Administrators: Pathway to Leadership Positions in the NCAA: A Qualitative Analysis, PhD Thesis, North Carolina State University, Raleigh, NC.

Further Reading

Acosta, V., Carpenter, L., 1985. Women in sport. In: Shu, D., Segrave, J., Becker., B.J. (Eds.), Sport and Higher Education, first ed. Human Kinetics Publishers, Champaign, IL, pp. 313—325.

Acosta, V., Carpenter, L., 1988. Perceived Causes of the Declining Representation of Women Leaders in Intercollegiate Sports—1988 Update. first ed. Brooklyn College, Brooklyn, NY, Unpublished manuscript.

Acosta, V., Carpenter, L., 2004. Women in Intercollegiate Sport. A Longitudinal Study— Twenty-Nine Year Update. first ed. Brooklyn College, Brooklyn, NY, Unpublished manuscript.

Burdman, P., 2002. Old problem, new solution? Can programs such as the NCAA's leadership institute for ethnic minority males boost the numbers of Black head coaches, athletic directors? Black Issues High. Educ. 19, 24—28.

Cuneen, J., 1988. A preparation model for NCAA Division I and II athletic administrators. In: North American Society for Sport Management.

Delano, L., 1988. Understanding Barriers That Women Face in Pursuing High School Athletic Administrative Positions: A Feminist Perspective. PhD Thesis, The University of Iowa, Iowa City, IA.

Delano, L., 1990. A time to plant—strategies to increase the number of women in athletic leadership positions. J. Phys. Educ. Recreat. Dance. 61, 53—55.

Dunning, E., 1999. Sport Matter: Sociological Studies of Sport, Violence, and Civilization. first ed. Rutledge, London.

NCAA., 2008. Ethnicity and gender demographics of NCAA member institutions' athletic personal.

Fitzgerald, M., Sagaria, M., Nelson, B., 1994. Career patterns of athletic directors: challenging the conventional wisdom. J. Sport Manage. 8, 14–16.

Hart, B., Hasbrook, C., Mathes, S., 1986. An examination of the reduction in the number of female interscholastic coaches. Res. Q. Exerc. Sport. 57, 68–77.

Hatfield, B., Wrenn, J., Bretting, M., 1987. Comparison of job responsibilities of intercollegiate athletic directors and professional sport general managers. J. Sport Manage. 1, 129–145.

Hay, R., 1986. A proposed sports management curriculum and related strategies. Paper presented at the meeting of the North American Society for Sport Management, Kent, OH.

Herron, L., 1969. A Srvey to Compare the Educational Preparation and Related Experiences and Selected Duties of Collegiate Athletic Directors. PhD Thesis, University of Utah, Salt Lake City, UT.

Landry, D., 1983. What makes a top college athletic director? Athletic Admin. 18, 20.

Myles, R., 2005. The Absence of Color in Athletic Administration at Division I Institutions. PhD Thesis, University of Pittsburgh, Pittsburgh, PA.

NCAA study on women in intercollegiate athletics: Perceived barriers of women in intercollegiate athletics careers, 1989. National Collegiate Athletic Association, 9.

Quarterman, J., 1992. Characteristics of athletic directors of historically black colleges and universities. J. Sport Manage. 6, 52–63.

Strengthening historically black colleges and universities program and strengthening historically graduate institutions program; final regulations, 1987. Federal Register. 52, pp. 30535–30543.

Terry, S., 1988. The private college athletic administrator. Athletic Administration: Official Publication of the National Association of Collegiate Directors of Athletics, 23, pp. 551–593.

Theberge, N., 1983. Towards a feminist alternative to sport as a male preserve. In: NASSS, St. Louis, MO.

Tiell, B., 2004. Career Paths, Roles, and Tasks of Senior Women Administrators in Intercollegiate Athletics. first ed. Pro Quest Information and Learning Company, Ann Arbor, MI.

Truiett-Theodorson, R., 2005. Career Patterns of African-American Athletic Directors at Predominantly White Higher Education Institutions: A Case Study. PhD Thesis, Morgan State University, Baltimore, MD.

Williams, J., Miller, D., 1983. Intercollegiate athletic administration: preparation patterns. Res. Q. Exerc. Sport. 54, 398–406.

INSTITUTIONAL AND SOCIETAL CULTURES

Institutional and Societal Cultures

Mary Oling-Sisay
Alliant International University, San Diego, CA

This chapter illustrates the current state of affairs, explores concepts of social exclusion–inclusion (Cox, 2001; Green, 2005; Ibarra, 1993; Meyerson and Fletcher, 2000) and organizational culture, and it posits that existing conceptual models do not align institutional practices with policies. The chapter also highlights some of the unintended consequences and tensions that can arise in the course of implementing strategies for a more inclusive institution and proposes a model to enhance diversity and inclusion.

A persistent issue facing today's workplace is that of inclusion and exclusion. Research is rife with agreement that people of color often find themselves excluded from networks of information and opportunity (Cox, 2001; Creed and Scully, 2000; Jackson and Jones, 2001; Meyerson and Fletcher, 2000). A common explanation is that ubiquitous perceptions and general sense of discomfort with those who are perceived as different can be the reason for their exclusion from important institutional processes and resources (Barak, 2000; Dass and Parker, 1999; Davidson, 1999; Ely and Thomas, 2001). It is not uncommon in faculty or administrative searches to hear comments that state that "the candidate is not a good fit." The definition of "fit" often is the notion that the individual may not think or act like the mainstream. In general, people tend to feel more comfortable with others with whom they share important characteristics, strengthening in-group/out-group perceptions and creating exclusionary behaviors (Blau, 1977).

Scholars of diversity, equity, and inclusion posit that of all human needs, the need to be seen, included, and accepted by others is perhaps one of the most powerful (Arredondo, 1996; Barbour, 2008; Betances et al.,

29

2006; Creed and Scully, 2000; Johnson, 2006). Inclusion can be viewed as a sense of feeling welcome, of acceptance, and of comfort as an integral part of an organization. The concept of inclusion—exclusion refers to an individual's sense of their position relative to the mainstream of an organization.

For women of color, this can be manifested in the formal processes like access to decision-making channels, and in the informal processes such as acknowledgment of one's presence, saga narrative, "water cooler sessions," and lunch or after hour's meetings where information and decisions are sometimes made (Barak, 2000; Farmer, 1993; Sennett, 2000). In some instances, the exclusion is blatant—typically with explanations such as "I did not think you needed to or wanted to be there."

In some cases, it is even more blatant wherein team environments members of the mainstream dominate the conversation and by so doing render the person of color "invisible" (Barak, 2000; Rolle et al., 2000; Trepagnier, 2006; Turner and Myers, 2000; Villalpando and Bernal, 2002; Watson, 2001). The other side of the coin is when the same "invisibly visible" individual suddenly becomes acutely visible when matters of diversity, especially race and ethnicity, is the subject of the discussion. If not appropriately addressed by leadership, the concept of invisibility and visibility engenders loss of interest, curtails innovation, and deprives the institution of creativity through the opportunity to engage all members with different perspectives and experiences (Cox, 2001; Ely and Thomas, 2001; Gregory, 2003; Hollis, 2007; Johnson, 2006).

Many researchers in diversity view the concept of inclusion (Barbour, 2008; Johnson, 2006; Kendall, 2006; Marcus, 2000) as central to diversity scholarship and practice. Social scientists have long utilized this approach to explain human preferences and behaviors. In the field of psychiatry, Schutz (1958, 1966) declared that a sense of affection, control, and inclusion are essential components for effective group interactions. He defines inclusion as:

> The need to establish and maintain a feeling of mutual interest with other people. This feeling includes (1) being able to take an interest in other people to a satisfactory degree and (2) having other people take interest in the self to a satisfactory degree. (*Schutz, 1958, p. 14, 18*).

Therefore, it is important for senior leadership in higher education to state clearly and unequivocally that equitable treatment of everyone is criteria for organizational effectiveness, excellence, and success (Stubblefield, 2005; Sullivan and Tuana, 2007; Touchton, 2008). Without this criteria, some will likely define excellence based on the needs of the dominant members rather than as a means of attaining quality and institutional progress through the capitalization of multiple perspectives and talents (Meyerson and Fletcher, 2000) (Figure 3.1).

FIGURE 3.1 Basic model of inclusion. *Source: Adapted from Mor Barak (1999). Copyright 1999 by M.E. Mor Barak. Reprinted by permission.*

THE ORGANIZATIONAL CULTURE CONTEXT

Schein (1987, 1992, 2004) declared that any successful change initiative must begin by addressing underlying assumptions and beliefs because merely changing the cultural artifacts and behaviors make it easier for the institution to revert to its old ways in a very short period of time. Culture comprises the values, beliefs, underlying assumptions, attitudes, and behaviors shared by a group of people. Culture is the behavior that results when a group arrives at a set of—generally unspoken and unwritten—rules for working together (Schein, 1988, 1996, 2004). Kuh and Whitt (1988) asserted that culture is

> the collective, mutually shaping patterns of norms, values, practices, beliefs, and transformations that guide the behavior of individuals and groups in an institute of higher education and provide a frame of reference within which to interpret the meaning of events and actions on and off campus. (pp. 12–13)

According to scholars Bolman and Deal (2003, 2008) and Schein (1992, 1996, 2004), culture is represented in an organization's language, decision making, symbols, stories and legends, and daily work practices. It is "the deeper level of basic assumptions and beliefs that are: learned responses to the group's problems of survival in its external environment and its problems of internal integration; are shared by members of an organization; that operate unconsciously; and that define in a basic 'taken-for-granted' fashion in an organization's view of itself and its environment" (Schein, 1988).

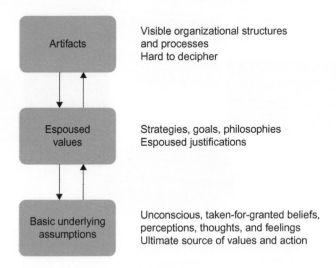

FIGURE 3.2   Uncovering the levels of culture. *Source: From Schein (1992). Copyright 1992 by Jossey-Bass. Reprinted by permission.*

Schein's (1988) seminal work on organizational culture provides a basis for understanding the complexity of organizational culture (Figure 3.2).

An organization's culture is made up of all of the life experiences each member brings into the organization. Culture is especially influenced by the organization's founder, executives, and other managerial staff because of their role in decision making and strategic direction (Schein, 2004). Organizations stay the same because it is the easier path for them to take. Certain practices can easily lead to "group think" (Cox, 2001) where different perspectives are avoided because it interferes with the cohesion of the dominant group. Scholars of organizational culture and diversity and inclusion postulate that many organizational cultures lend credence to this notion by promoting the notion of the "model employee." Typically, this individual is conversant with the dominant culture.

Those individuals who do not fit into this mold are ignored, excluded, and by so doing marginalized (Meyerson and Fletcher, 2000). Women of color often find themselves in this predicament. This is especially crucial as they endeavor to maintain their individual identity. This personal identity provides them with coping skills and a sense of individual autonomy as they navigate their sense of place within formal and informal encounters because of their race, ethnicity, and gender (Cox, 2001; Ramey, 1995; Watson, 2001).

Studies of microcultures and their attendant hidden rules in organizations and the "normal way of doing business" have generated numerous strategies that have enhanced the workplace. More "excavation" is needed to further uncover subtle concepts and patterns that foster organizational cultures. These cultures effectively aid the challenge of transformational change in creating an institutional environment that is inclusive. As mentioned before, an organization's culture consists of the values and norms established by the institution as well as personal attitudes, behaviors, and experiences that employees themselves bring to their positions.

However, all too often the emphasis on diversity in higher education tends to focus on compositional diversity without addressing issues of inclusion—exclusion, organizational culture, and climate for cultural interactions and communications. It is important to recognize retention as the other side of the recruitment coin. Although beneficial and needed to enhance the climate and access for underrepresented members, diversity programs that ignore retentive strategies can also serve to marginalize people of color (Appadurai, 2009; Brickson, 2000). Farmer (1993):

Faculty and administrators are located somewhat differently within academic hierarchy; however, people of color, whether faculty or administrators, are "commoditized" in ways that their White counterparts are not (Johnson, 2006; Sullivan, 2006; Sullivan and Tuana, 2007). This "commoditization" especially occurs when, through either internal or external pressures, educational institutions are forced to diversify their personnel. What often results is the positioning of faculty and administrators so that they have visibility (thus improving the institution's public image). However, these administrators have very little autonomy or power (and power remains concentrated with the White race). Under these circumstances, people of color, much like the scholarship about people of color, are given place but not importance.

RECONCEPTUALIZING INCLUSIVITY

While many institutions express their diversity commitment in their mission statements and strategic planning, these very institutions also demonstrate a dearth of programs and practices that result in a diverse learning environment (Bensimon, 2004; Milem and Hakuta, 2000).

Visions, mission statements, and memoranda are not sufficient to utilize concepts of diversity and inclusion as transformational change initiatives. Higher education leadership must understand and utilize techniques to imbibe in all members of their institutions, values and norms that transcend the macroculture and cross microcultures (Brickson, 2000; Brown, 2000). This approach provides opportunities to

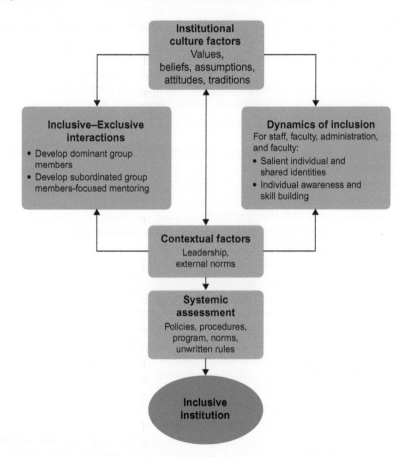

FIGURE 3.3   Reconceptualized model for institutional inclusion to enhance the advancement of female administrators of color.

address cultural ambiguities that negatively affect diversity and inclusion. Reconceptualizing inclusivity and enhancing the experiences of female administrators of color must begin with acknowledgment of the role institutional culture plays. Culture is instrumental in inclusive–exclusive interactions as well as power relations (Schein, 1992) (Figure 3.3).

To increase diversity and inclusivity in the ranks of higher education administration, institutions must begin by recruiting increased numbers of people of color. It is through increasing diversity in the leadership pipeline and through professional development that diversity will permeate all levels of administration. Further, institutions must be explicit in their processes, practices, and communications regarding issues of race, gender, and organizational culture. In addition, institutions must be willing to include a system of rewards and consequences for championing and practicing inclusivity.

Existing scholarship on women's participation in senior leadership in higher education indicates that current practices to recruit, advance, and sustain women of color in senior leadership are inadequate (June, 2007; Mcshane and Von Glinow, 2005; Smith, 2000; Watson, 2001). Research has begun to emerge on the experiences of people of color in administrative positions in higher education. However, much of the research is focused on the presidency. While this is no doubt an important agenda, it is critical to examine what is happening to the other leadership ranks. For instance it is clear that most of the people of color in senior leadership are clustered in certain positions, namely, chief diversity officers and chief student affairs officer (ACE, 2008; Mcshane and Von Glinow, 2005; Schmidt, 2008). Yet these are positions with "glass ceiling" tendencies.

Further, institutions need to place special attention on searches at the dean and vice presidents level to ensure the applicant pool is diverse. Currently, people of color are experiencing a decline in representation in these two executive-level positions (ACE, 2008). It is particularly important to address this decline because barring changes in pathways to the presidency, these are positions that traditionally groom incumbents to move to the college presidency. What is needed is a top—down and bottom—up engagement. There are many factors and conditions influencing and determining the advancement of women of color in higher education (Holmes, 2003; Jogulu and Wood, 2006). Clearly, this is not only dependent on the personality, identities, experiences, and backgrounds of individuals with external circumstances that institutions cannot change. Rather, it is also a factor of conditions created by the institution that can be improved, such as policies and procedure, professional development, and organizational culture and climate (Bolman and Deal, 2003, 2008; Cameron and Quinn, 2006; Barbour, 2008).

References

American Council on Education (ACE), 2008b. The American College President Study: Key Findings and Takeaways. Author (ACE), 2008b, Washington, DC.

Appadurai, A., 2009. Higher education's coming leadership crisis. Chron. High. Educ. 55 (31).

Arredondo, P., 1996. Successful Diversity Management Initiatives. first ed. Sage Publications, Thousand Oaks, CA.

Barak, M.E., 2000. The inclusive workplace: an ecosystems approach to diversity management. Soc. Work. 45, 339—353.

Barbour, J., 2008. Organizational culture and institutional discrimination. In: Western Political Science Association, Manchester Hyatt, San Diego, CA.

Bensimon, E.M., 2004. The diversity scorecard: a learning approach to institutional change. Change. 36 (1), 45–52.

Betances, S., Torres, L., Souder, L., 2006. The Business Case for Diversity. first ed. Souder, Betances & Associates, Inc, Chicago, IL.

Blau, P.M., 1977. Inequality and Heterogeneity: A Primitive Theory of Social Structure. Free Press, New York, NY.

Bolman, L., Deal, T., 2003. Reframing Organizations. third ed. Jossey-Bass, San Francisco, CA.

Bolman, L., Deal, T., 2008. Reframing Organizations. fourth ed. Jossey-Bass, San Francisco, CA.

Brickson, S., 2000. The impact of identity orientation on individual and organizational outcomes in demographically diverse settings. Acad. Manage. Rev. 25, 82–101.

Brown, M., 2000. Organization & Governance in Higher Education. fifth ed. Pearson Custom Publishing, Boston, MA.

Cameron, K., Quinn, R., 2006. Diagnosing and Changing Organizational Culture. first ed. Jossey-Bass, San Francisco, CA.

Cox, T., 2001. Creating the Multicultural Organization. first ed. Jossey-Bass, San Francisco, CA.

Creed, W., Scully, M., 2000. Songs of ourselves employees' deployment of social identity in workplace encounters. J. Manage. Inq. 9, 391–412.

Dass, P., Parker, B., 1999. Strategies for managing human resource diversity: from resistance to learning. Acad. Manage. Exec. 13, 68–80.

Davidson, M., 1999. The value of being included: an examination of diversity change initiatives in organizations. Perform. Improv. Q. 12, 164–180.

Ely, R., Thomas, D., 2001. Cultural diversity at work: the effects of diversity perspectives on work group processes and outcomes. Admin. Sci. Q. 46, 229–273.

Farmer, S., 1993. Place but not importance: the race for inclusion in academy, Spirit, Space & Survival. first ed. Routledge, New York, NY, pp. 196–217.

Gregory, R., 2003. Women and Workplace Discrimination. first ed. Rutgers University Press, New Brunswick, NJ.

Hollis, R., 2007. Leader-as-teacher: a model for executive development success. Organ. Dev. J. 25.

Holmes, S., 2003. Black female administrators speak out: narratives on race and gender in higher education. NASAP J. 6, 47–68.

Ibarra, H., 1993. Personal networks of women and minorities in management: a conceptual framework. Acad. Manage. Rev. January, 56–87.

Jackson, J., Jones, L., 2001. A new test for diversity: retaining African American administrators at predominantly white institutions, Retaining African Americans in Higher Education. first ed. Stylus, Sterling, VA.

Jogulu, U., Wood, G., 2006. The role of leadership theory in raising the profile of women in management. Equal Opportun. Int. 25, 236–250.

Johnson, A., 2006. Privilege, Power, and Difference. first ed. McGraw-Hill, Boston, MA.

June, A., 2007. Presidents: same look, different decade. Chron. High. Educ. 5, 33.

Kendall, F., 2006. Understanding White Privilege. first ed. Rutledge, New York, NY.

Kuh, G.D., Whitt, E.J., 1988. The Invisible Tapestry: Culture in American Colleges and Universities. Association for the Study of Higher Education, Washington, DC, ASHE-ERIC Higher Education Report No. 1.

Marcus, L., 2000. Staff diversity and the leadership challenge. Equity Excel. Educ. 33, 61–67.

McShane, S., Von Glinow, M., 2005. Organizational Behavior. first ed. McGraw-Hill, Boston, MA.

Meyerson, D., Fletcher, J., 2000. A modest manifesto for shattering the glass ceiling. Harv. Bus. Rev. 78 (1), 126–136.

Milem, J.F., Hakuta, K., 2000. The benefits of racial and ethnic diversity in higher education. Minorities in Higher Education: Seventeenth Annual Status Report. American Council on Education, Washington, DC, pp. 39–67.

Mor Barak, M.E., 1999. Beyond affirmative action: toward a model of diversity and organizational inclusion. Admin. Soc. Work. 23, 47−68.

Ramey, F., 1995. Obstacles faced by African American women administrators in higher education: how they cope. West. J. Black Stud. 19, 113−119.

Rolle, K., Davies, T., Banning, J., 2000. African-American administrators' experiences in predominantly white colleges, universities, and community college. J. Res. Pract. 24, 79−94.

Schein, E., 1987. Organizational Culture and Leadership. first ed. Jossey-Bass, San Francisco, CA.

Schein, E., 1992. Organizational Culture and Leadership. second ed. Jossey-Bass, San Francisco, CA.

Schein, E., 2004. Organizational Culture and Leadership. third ed. Jossey-Bass, San Francisco, CA.

Schmidt, P., 2008. Colleges have blind spots in presidential searches. Chron. High. Educ. 54, A64.

Schutz, W.C., 1958. FIRO: A Three-Dimensional Theory of Interpersonal Behavior. Holt, Rinehart & Winston, New York, NY.

Schutz, W.C., 1966. The Interpersonal Underworld. first ed. Science & Behavior Books, Palo Alto, CA.

Sennett, R., 2000. Work and social inclusion, Social Inclusion. first ed. St. Martin's Press, New York, NY.

Smith, D., 2000. Women at Work. first ed. Prentice Hall, Upper Saddle River, NJ.

Stubblefield, A., 2005. Ethics along the Color Line. first ed. Cornell University Press, Ithaca, NY.

Sullivan, S., 2006. Revealing Whiteness. first ed. Indiana University Press, Bloomington, IN.

Sullivan, S., Tuana, N., 2007. Race and Epistemologies of Ignorance. first ed. State University of New York Press, Albany, NY.

Touchton, J., 2008. A Measure of Equity: Women's Progress in Higher Education. first ed. Association of American Colleges and Universities, Washington, DC.

Trepagnier, B., 2006. Silent Racism. first ed. Paradigm Publishers, Boulder, CO.

Turner, C., Myers, S., 2000. Faculty of Color in Academe. first ed. Allyn & Bacon, Boston, MA.

Villalpando, O., Bernal, D., 2002. A critical race theory analysis of barriers that impede the success of faculty of color, The Racial Crisis in American Higher Education. first ed. State University of New York Press, Albany, NY, pp. 243−269.

Watson, L., 2001. In their voices: a glimpse of African-American women administrators in higher education. Natl. Assoc. Stud. Affairs Prof. J. 4, 7−16.

B. INSTITUTIONAL AND SOCIETAL CULTURES

CAREER PATHWAYS

A Tale of Two Professors

Kimberly M. Coleman[1] and Kera Z. Watkins[2]

[1]College of Public Health, Georgia Southern University, Statesboro, GA
[2]Department of Computer Science, Georgia Southern University,
Statesboro, GA

It was the best of times, it was the worst of times, it was the age of wisdom, it was the age of foolishness, it was the epoch of belief, it was the epoch of incredulity, it was the season of Light, it was the season of Darkness, it was the spring of hope, it was the winter of despair, we had everything before us, we had nothing before us... **(Charles Dickens, 1859)**

Two African-American girls linked by genetics as first cousins, whose paths would connect them professionally as members of the academy at the same university at the same time. This is not, however, a Dickens tale. On the contrary, this story is true and it chronicles their journey from the interview through the pretenure review process. It vividly illustrates the challenges and barriers each confronted as the lone African-American female, junior professor in their respective departments and the lessons learned along the way (Figure 4.1).

It has been almost 4 years since we started our full-time positions as assistant professors, and we still question how we landed in this place called academia. How we, first-generation family members, earned doctorates in the same year. How we were hired in two different colleges at the same university is a strange phenomenon. Even more phenomenal are the parallel experiences we have endured during these years as tenure-track faculty members.

The impact of these parallel experiences has altered our views of the professoriate and our decisions to leave or remain in academia.

Kimberly Coleman and Kera (Bell) Watkins were born 9 months, almost to the day, apart. Our mother and father are siblings; due to our close ages and parentage, we grew together like siblings also. We played,

Navigating Academia: A Guide for Women and Minority STEM Faculty.
DOI: http://dx.doi.org/10.1016/B978-0-12-801984-9.00004-3.

FIGURE 4.1   Kera and Kim, Washington, DC, 1978.

fought, were mischievous, endured correction and rebuke, were nurtured spiritually and babysat by our maternal grandmother, Irma Lee Bell, together. We also had something else in common—mothers who believed in our greatness and who always told us that we could do anything we wanted to do.

Kim was lazy about homework by the time she began the seventh grade. Kim remembers a night where she was fighting her mother about completing her homework. Her mother, finally having reached her boiling point, looked her daughter and said, "Kim, I don't care if you sit here all night. I'm not standing over you to make sure you finish this. You are of at least average intelligence. Therefore, you can do anything you want if you put your mind to it. I already have my diploma and degrees. You do what you want. I'm going to bed." It was clear that she wasn't coming back and Kim eventually completed the assignment. At that time, her mother's words were irritants to her teenage wisdom. As Kim got older, she believed her mom's wisdom more.

Kera believes that education is like a skeleton key. It can be used to open any door... even ones that might be slammed in her face. Kera eventually became one of those kids who wanted to be a lawyer, a doctor, a business owner, an astronaut, a Girl Scout leader, and a musician all wrapped up into one! Kera really didn't see a reason why she couldn't do it!

For all of our similarities, we were academic opposites as children and still are as adults. The daughter of a speech and language pathologist, Kim loved public speaking, reading, and writing. Kim was drawn to those subjects where those skills were challenged and further developed. Kera grew up adoring mathematics, especially calculus in high school. Kera's mom who is a registered nurse, inspired her.

We excelled in those subjects that focused on our intellectual strengths and interests. Following high school, we applied to, attended, and earned our undergraduate degrees from Spelman College in Atlanta, GA. Spelman College, established in 1881, is a nationally-ranked, historically Black college for women and dedicated to the development of the "intellectual, ethical, and leadership potential of its students" (The Spelman College Office of Alumnae Affairs, 2010).

Kim majored in psychology, driven by an intellectual curiosity to understand human behavior. Kera found that her love for mathematics guided her to pursue a Bachelor of Science degree in this discipline. Stereotyped rumors claiming that historically Black colleges and universities are less rigorous were definitely not true of the Spelman experience. The faculty and staff were supportive, but students learned more than coursework. We also learned about life— we were taught about the challenges of being educated African-American women. Each of us excelled, graduating with honors in our respective fields, and determined to further our educational pursuits in graduate school.

Kim and Kera graduated from Spelman in 1994 and 1995, respectively. We went our separate ways, eventually earning graduate degrees. Kim earned a Master of Public Health degree in Health Behavior and Health Education from the University of Michigan and a Doctor of Philosophy degree in Health Education from Southern Illinois University Carbondale in 2006. Kera earned a Master of Science in Computer Science from Clark Atlanta University, and a Doctor of Philosophy degree in Computer Science from North Carolina State University in 2006 also. Kim has worked at a number of places over the years. She has taught, managed, directed, coordinated, and conducted research.

Kera notes that regardless of what she believed about herself, she encountered people who assumed that she was incapable of something even before getting to know anything about her. For example, Kera distinctly recalls during her college years signing up for an electronics class that really excited her. Kera arrived in class the first day of the semester close to 10 min early and sat in the front row. The instructor for the class was already at the front of the room writing something on the board. A bell rang and the instructor stopped writing and turned to scan the class of 30—40 students for several seconds. Afterward, he turned back toward the board to continue writing. While he was writing and still had his back to the class, he announced that he had never given a woman an A.

Kera's excitement quickly morphed into insecurity and horror. All of a sudden, she felt like all eyes were on her. Kera cautiously turned to see the other students in the room, and she realized that she was the only woman in the room. Kera assumed the instructor was talking to her.

The rest of that first class is a complete blur. Kera can't remember if she even took notes. She remembers quickly dropping the class by the end of the day, but she always wondered about the instructor's rationale. Had a woman never earned an A in his class? What percentage of the class made A's or any passing grades in his class? Did he believe that women should not even take his course or maybe even major in his particular technical field? Did he automatically (or even subconsciously) lower a person's grade if the student was a woman? Kera realizes that she will never know the answers to any of these questions in the case of that instructor.

Kera made an important discovery about herself as a person during her college years. There are some who will try to sabotage her. There are some who will try to lift her up. There are some people who will not care either way about her. There are some who will be scared of her. There are some who will be scared for her.

Based on estimations shown in the National Science Foundation's *Women, Minorities, and Persons with Disabilities in Science and Engineering* (2003), roughly 4% of the nation are computing professionals and 8% are health professionals possessing doctorates. The statistics focused on African-American women reveal the underrepresentation of this demographic group. Kera is included in the 0.5% of African-American women computing professionals holding a doctorate in the United States, while Kim is part of the 10% of African-American women health professionals holding a doctorate in the United States.

In November of 1999, the Carnegie Foundation for the Advancement of Teaching expanded its Classifications of Institutions of Higher Education to reflect a range rather than a single typology (Carnegie Academy, 1999). As a result of this change, our university engaged in a mission and vision-based paradigm shift as it sought to clarify the practical definition of the Doctoral Research University (DRU) classification.

This section of the chapter details how the lack of clarity among the university administration, faculty, staff, and students affected their experiences. Most notably, the section discusses how people strived to balance teaching, research, and service in the shadow of the university's DRU identity crisis. Both women were prepared during their interview with the "right" questions directed to the correct administrator (i.e., dean of the college, department chair), took copious notes of their answers, and clearly expressed their needs and concerns. However, the reality was that once they signed the contract, many of the agreements and promises were not fulfilled.

Initially, they believed academia to be a liberal environment and that faculty members, regardless of gender or ethnicity, regularly engaged in

FIGURE 4.2   Dr. Kera Z. Watkins and Dr. Kimberly M. Coleman (Statesboro, GA, Fall 2010).

"outside the box" inquiry. They thought that academia valued and sought to understand cultural and ethnic differences. They thought that tenured members of the academy would support junior faculty, especially those from underrepresented groups, because they valued sustainable diversity and multiculturalism within the academy. However, their actual experiences included isolation and an understanding that some of their ideals were not realistic. The naïve idealism and optimism with which they entered academia is not unique. However, the benefit of their acknowledgment can introduce those considering, entering, or remaining established members of academia to some of the unexpected truths (Figure 4.2).

References

Carnegie Academy for the Scholarship of Teaching and Learning, 1999. Informational Program. Booklet. Carnegie Foundation for the Advancement of Teaching, Menlo Park, CA.
Carnegie Foundation for the Advancement of Teaching, 1994. A Classification of Institutions of Higher Education, 1994 Edition. Princeton:Author.

Dickens, C., 1993. A Tale of Two Cities. Wordsworth Editions Ltd, London.
National Science Foundation, 2003. Women, Minorities, and Persons with Disabilities in Science and Engineering (NSF 03-207). Author, Arlington, VA.
The Spelman College Office of Alumnae Affairs, History and Traditions Reference Guide, Web. 28 October 2010. Available from: <http://www.spelman. edu/about_us/news/ publications.shtml#historytraditions>.

A Career in Science

Raquel Diaz-Sprague
Technical Support, Inc., Columbus, OH

"Science is a girl thing" she says.

There are dozens of programs for Women in Science or Girls in Science at institutions around the nation. But, if you search the Internet for "women in science day" the first three entries will lead to the Women in Science Day Program directed by Dr. Raquel Diaz-Sprague in Columbus, OH, at http://awisco.osu.edu/.

EARLY FORMATIVE YEARS (1947—1963)

Raquel was born in the third largest city of Peru, Trujillo, which is the capital of La Libertad region. Trujillo is a highly cultured and commercially active coastal city located in northwestern Peru some 500 km north of the capital city of Lima. Her father was Juan Diaz Silva (born 1917), from the Andean area of Cascas, and her mother was Maria Alejandrina Ramirez de Diaz (born 1914), from the small town of Huaral near Lima.

Her father who was the oldest living of six children left his native Cascas to attend the prestigious San Juan High School in Trujillo. After he graduated from high school, he attended the Universidad Nacional de Trujillo (UNT) where he obtained a Bachelor of Science in Economics and Accounting, and he later received certification as a Certified Public Accountant. He loved to play the guitar and mandolin and was an accomplished swimmer and marathon runner. He worked as a professional accountant his entire life, and he also taught for a while at the UNT. His work put him in contact with English-speaking people and he was an avid learner of the English language.

Her mother, Maria, who had not been able to finish high school due to lack of financial resources was an avid reader and life-long learner.

Navigating Academia: A Guide for Women and Minority STEM Faculty.
DOI: http://dx.doi.org/10.1016/B978-0-12-801984-9.00005-5.

Prior to marriage, she had worked as an accounting assistant. When her children were of school age, she attended night school for several years and obtained certificates in Crafts, Culinary Arts, and Clothing Design. She was an instructor in a vocational school, and later she established a private academy for culinary arts and crafts. Juan and Maria were married in October 1945 and had four children, two girls and two boys, born between 1946 and 1953. Raquel was born on October 10, 1947.

Raquel and her siblings were raised in a loving family that stressed advanced education, healthy lifestyles, and service. Raquel and her sister attended private Catholic schools run by nuns, while their brothers attended paramilitary public schools. Raquel excelled in school since kindergarten. She began reading newspapers by age 4, was considered precocious by her teachers, and was quite outspoken.

The family lived in a long, narrow 150-year old home built of thick adobe bricks, which have resisted the test of time, including several strong earthquakes. The family of eight included two great aunts, Carmelita and Santos, who had helped raise Maria and her sister, Albina, since they were kids. In her great aunt Carmelita, a dedicated professional midwife, Raquel saw a role model of a health professional devoted to service.

From an early age, Raquel aspired to become a health professional. She took a rigorous math and science curriculum in high school, and she was also fascinated by the arts and literature. She read and wrote poetry. During the summer, the family visited nearby beaches and often traveled to Cascas, province of Gran Chimu, the paternal homeland. Her paternal grandparents owned perhaps the largest private home in Cascas. Raquel and her sister liked to play in the many rooms of that enormous house that had 36 doors.

Her grandmother, Dona Teodolinda, ran a successful bakery and her grandfather, Don Manuel Diaz Murrugarra, had a distinguished career in public service. He served for decades first as a Councilman, City Treasurer, and then Mayor. He is remembered for his tireless work to bring development and educational opportunities for the citizens of Cascas. Because of his many accomplishments, the city's first national high school "Gran Unidad Escolar Manuel Diaz Murrugarra" was named in his honor.

All of Raquel's paternal uncles and aunts attended universities and became successful professionals—accountant, physician, dentist, pharmacist, teacher, and pilot. One of her uncles Manuel, the dentist, ran for political office, (unsuccessfully) a couple of times. Her youngest brother, Jaime (born 1953) became a sociologist and instructor at the Universidad de los Andes in Merida, Venezuela. Her brother Johnny (born 1950) is a mathematics teacher in Trujillo and her older sister Gloria Jenny (born 1946) is a ballet teacher in Machala, Ecuador.

Raquel was a stellar student in the parochial school, Colegio Hermanos Blanco, which she attended from Grade 1 to Grade 7. She was honored with the First Prize for Outstanding Achievement and Exemplary Behavior in each of those years. She was often chosen to deliver speeches for religious events representing her class to the school.

She was a fast and avid reader of her father's book collection and an old encyclopedia. She also read about the lives of the saints from the school library. She was fascinated by the mystical appeal and infinite compassion of San Martin of Porres and Saint Francis of Assisi, and she was intrigued by the intellectual life of Sor Juana Inez de la Cruz. For a while, she considered entering a convent after high school to become a nun and pursue a life of pure intellectual and spiritual development. However, in time she grew disenchanted with nuns by certain unkind attitudes of the Superior Mother of her school.

She and her sister left Colegio Hermanos Blanco after Grade 7 and they attended Grade 8 at a private school. They enrolled in Colegio San Vicente de Paul for Grade 9 and 10. Raquel flourished academically and socially in Colegio San Vicente de Paul. Her most notable accomplishment was winning a writing contest, in English, in which the top students from 14 schools competed. Raquel's picture appeared in the front page of the local newspaper "La Industria." Her school honored her. She gave a speech for her high school graduation in December 1963 and she was also recognized in her community as a "campeona."

COLLEGE LIFE (1964–1971)

Following her aspiration to become a health professional, Raquel applied to the prestigious UNT in 1964. The university entrance exam is extremely competitive. About 10,000 students vied for 1,000 spaces. The top 100 or so scorers are admitted to premedical school or pre-pharmacy school. Raquel's score was high enough for pre-pharmacy school, but only the top 60 students (many of whom were taking the test for the second or third time) gained admission to the premedical school.

Rather than wait a year to retake the test, Raquel declared Pharmacy as her intended major and began taking courses in basic sciences in pre-pharmacy at the UNT. She excelled in her coursework and earned a place in the Pharmacy school along with 46 male and 3 female classmates. She received high grades in all her classes and after her first year, she earned a coveted Teaching Assistantship (TA) for the remaining of her student career. Her duties as a TA involved the preparation of laboratory reagents, culture media, slides, and so on. She also helped grade tests, had extended library privileges, and was paid a monthly stipend.

C. CAREER PATHWAYS

During her first year at the university, she also took evening classes in classic ballet and theater at the prestigious Casa de la Cultura, which was located close to the university. Raquel was well regarded for her philosophical insights and poetic language. She was also popular among her peers. Her pre-pharmacy class elected her the Queen of Pre-Pharmacy in 1965. In her first year as a Pharmacy student, she was elected Queen of the Freshman Class and later in a general election among students of all the other classes, she was elected Queen of the Facultad de Farmacia. This latest election was a very high honor. She was invited to participate in Faculty Council meetings representing students. She wore a tiara, and a red sash marked "Raquel I" and was feted at a variety of school events for an entire week. Raquel's parents attended to watch her.

Raquel was a poetry buff and dedicated book worm. She did not spend much time in sports, dances, or other social pursuits. One of her classmates, Olga, was her closest friend, confidant, and book discussion partner as they both read the same books. Her first boyfriend was an award-winning poet and painter of Polish ancestry. He dedicated a book to her and wrote several poems about her.

After graduating with top honors from the Facultad de Farmacia, she did the required 1,000 h of practice as a pharmacist at the Hospital Regional Docente de Trujillo. Additionally, on her own initiative, she took up a service project entitled "What is for breakfast in a slum in Trujillo?" For 3 months, Raquel traveled early in the morning to one of Trujillo's slum areas, Florencia de Mora. She visited 100 homes and interviewed the women about the food the family ate for breakfast. She took samples of the foods and analyzed them in the laboratory for nutritional value. The results showed that foods typically consumed by poor families were low in protein and nutrients and high in sugar. In her written report, she made recommendations to local health authorities to promote greater use of quinoa, a high-protein cereal native to Peru.

To obtain her professional degree—Quimico Farmaceutico—which is a terminal degree and the equivalent of a PharmD in the United States, Raquel had to do a research thesis. Under the leadership of Microbiology Professor, Dr. Jesus Garcia Alvarado, she did a 24-month research study entitled "Determination of the Sensitivity of 50 strains of *Pseudomonas aeruginosa* to Carbencillin." The motive for the research was Raquel's concern for the high rate of infants who died due to diarrhea. When she practiced pharmacy at the Hospital Regional de Trujillo, she had talked with many anxious and grieving mothers about the severe (often greenish) diarrhea affecting their infants. She obtained hundreds of samples and isolated one of the bacteria implicated. She isolated 50 unique strains of *Pseudomonas aeruginosa*, the blue-green pigmented, often deadly bacterium. She then tested the bacteria against numerous antibiotics and combinations of antibiotics, with a special emphasis in Carbenicillin. Her friend and classmate, Olga, tested the effects of Gentamicin on the 50 *Pseudomonas* strains.

In August 1971, Raquel defended her thesis and received the title of Bachelor of Science in Pharmacy and Biochemistry, plus the professional title of Quimico-Farmaceutica, and she was admitted to the Colegio Farmaceutico del Peru. In September 1971, Raquel started a job teaching biology and chemistry at Escuela Normal, a college for science teachers.

Raquel's immediate goal was to pursue graduate studies in microbiology in the United States. She entered in national contests for the LASPAU scholarship and Fulbright-Hays scholarship. In January 1972, she was awarded a LASPAU scholarship and was selected as a finalist from northern Peru for the prestigious Fulbright Scholarship. After a second round of standardized tests, GRE and TOEFL, essays, and interviews, in April she was notified of being awarded a Fulbright Fellowship to start graduate studies in the United States in fall 1972.

TRAVEL TO NORTH AMERICA (1972–PRESENT)

Winning a Fulbright Scholarship. Leads to Graduate Work in Microbiology at Ohio State University

Raquel was accepted as a graduate student in Microbiology at three research universities: one in Florida, one in Massachusetts, and one in Ohio. She chose to attend the Ohio State University (OSU) because it had the largest microbiology department and it offered an extensive array of courses, including food microbiology and fermentations.

Her trip to the United States was exciting. She received a generous travel allowance and two first class round-trip tickets to Lima. Her mother was able to go with her to Lima. A lot of her relatives, friends, and one UNT professor came to the Lima airport to see her off. Raquel's picture was in an article in the local newspaper about her having obtained a Fulbright Fellowship. Arriving into the Miami International Airport was almost surreal. She was whisked through customs and treated as a diplomat. She was met by an official of the US Department of State, who officially welcomed her to the country as an Exchange Visitor. Raquel was absolutely astounded by the massive airport and great number of planes.

Arriving into Columbus, OH, Raquel was met by members of the International Student Office and taken to an undergraduate dormitory because the graduate dorm was closed.

Raquel's first friends were her Chinese roommates. Together they experienced going to the nearby grocery store and to a department store. With a dictionary in hand, they tried to make purchases and make themselves understood. They succeeded about half the time.

Camp Akita

The Office of International Students organized a 3-day camp for all the international students. It was a really nice introduction to American culture, the love of the outdoors, the games, and food. Raquel enjoyed it immensely.

Graduate Study and Research

Raquel's goal was to study food microbiology to find ways to increase the nutritional value of foods commonly eaten in Peru. She investigated a process to make a "peanut beer," that is, a nutritious fermented beverage containing 3% alcohol made from peanuts, yeast, and sugar. She also worked on processes for making single-cell protein from yeasts leftover from brewery. She hoped to apply this knowledge when she returned to Trujillo.

Sexism and Harassment

The dark side of her graduate student days was rampant sexism and discrimination in the university, and sexual harassment from Raquel's adviser, Dr. K, an associate professor. Although Dr. K was helpful and supportive of Raquel's ideas for research, he behaved inappropriately toward her when the two were in his office. This behavior offended Raquel deeply.

She was embarrassed and uncomfortable by the professor's inquiries about her private life and his unwelcome advances. Several times he told her she should date, F, another graduate student who was single. "He is interested in you and I can see why," he would say. She felt pressured and started avoiding meeting Dr. K in his office, because he usually closed the door when she came up to ask him a question. She was also uncomfortable in F's presence.

She felt powerless—where could a first-year graduate student go to report being harassed at a laboratory? There was no official sexual harassment policy in place at the university. In Peru, where there had been very few women students, she had been one of four women in her pharmacy class of 50. She had never been subjected to this type of harassment. Was it her fault? She worried and prayed. After talking with a close friend and with her priest, she decided to get out of that laboratory.

She felt sad that she had to come to Columbus to be trained and do research that could help improve lives back home as well as enable her to support her parents and be of service to her community. Yet, the person most responsible for training her and helping her advance her research goals was actually preventing her from concentrating on her work.

When she had been a teaching assistant at UNT, she was used to working 12 or more hours a day in the biochemistry, microbiology, and pharmacology laboratories. But now, in the laboratory of Dr. K she felt unsafe working after hours. She often left the lab without completing her experiments to avoid being alone with either F or Dr. K.

The professor usually left the door of his office open with lights on, meaning he could return at any minute. Toward the end of 1972–1973, she requested a transfer out of that laboratory for "personal reasons," which was one of the few options she could check in a form. Her request was granted, but in the letter informing her of the transfer, the department chair cited "her inability to progress in research or to get along in the laboratory."

Not all was bleak. Raquel's closest friends at OSU were other foreign students, from all over the world, particularly Mai, from Thailand. Among foreign students, she found kinship and shared "otherness." She was an elected officer in the International Student Association, and she attended a lot of international shows, dinners, and events. Raquel worked closely with other student leaders and staff to make more services available to foreign students.

Transfer to Food Microbiology Lab

For her second year, Raquel was transferred to the laboratory of Dr. B, a full professor. Raquel had requested to work for him because he taught food microbiology. She was not aware that Dr. B was near retirement and that he did little or no research at the time. While he did not harass Raquel sexually, he disparaged her ideas. At their first meeting, Dr. B said that her research topic was a "waste of time and laboratory resources" because "no one will ever care for single-cell proteins or peanut beer." Often, he made jokes at her expense including telling others "Raquel is from Peru—where the narcos come from."

These slights made Raquel's determination stronger. She would find a way to do whatever it took to get her degree and get out of there. She vowed that once out of school, she would find a way to expose and combat the abuses and indignities that women had to endure from men in authority to persevere in the fields of science.

In Dr. B's lab, Raquel worked very hard, staying late into the night at the laboratory and on weekends. She made a novel discovery—the effect of microwave radiation in the induction of bacteriophages in *Bacillus cereus*. This finding resulted in a presentation at a microbiology conference in Venezuela and in a published abstract.

Dr. B did not care much. He did little to support Raquel's work. He generally did not spend time talking with his students. He spent a lot of his time in his office with the door closed and radio on. On her first day, Dr. B told Raquel, "You can work in this lab, but only for one year to get your Master's. I am near retirement. You should not waste time here, go home and get married. You are not PhD material."

Raquel was sad to hear that opinion. She had in fact assembled a bibliography and was testing her ideas and getting good results. She wanted to have an open future as everybody else. She did not appreciate being told she could not continue her research. Inexplicably and unjustly, Dr. B gave Raquel a Grade C in his course: Food Microbiology. He later told her that he would not support her admission to a PhD program in his laboratory and that she should not apply.

Raquel prayed and decided to apply anyway knowing that her chances of being approved were slim to none. In doing so, she was encouraged by the support of another Professor, Dr. C, an assistant professor in the Food Science Department with whom she had taken a course on microbial fermentations and gotten an A.

Dr. C was very interested in Raquel's research ideas and became a member of her committee. After she passed a comprehensive written and oral exam with flying colors, the committee deliberated about her fate. Professor C argued strongly to support Raquel for the PhD. He spoke about how exciting Raquel's ideas and work were, and he offered to be a mentor to her if Dr. B couldn't. Regrettably, Dr. C had no grant funding to support her, and Dr. B would not let her do the work in his lab. Nor did he favor that she become a TA. He said she didn't speak English well enough to teach and he didn't want responsibility for another student. He already had two students who were in their third and fourth year of a PhD program. He was waiting for them to finish so that he could retire.

The committee concluded there was no place for Raquel to continue doing research. Ironically, 3 years later, Raquel and Dr. B spoke at a meeting. Dr. B told her that the two male students had left his lab without graduating. He wistfully added, "I should have taken a chance on you."

After receiving her MS, Raquel took courses, applied for a grant, and traveled to Caracas to the Fourth Pan American Microbiology Congress. She presented her findings about the effect of microwave radiation on spores of bacteria. She was invited to apply for a research position in the Instituto Venezolano de Investigacion Cientifica.

She then went to Trujillo to spend Christmas with her family. She also had a meeting at the UNT where, as agreed with the Fulbright Commission, she was offered a faculty position in the Microbiology Department for the following academic year. As the funding was already in place, she was told she could start immediately. She would be a pioneer of sorts, the first female faculty in the Microbiology Department where she could perhaps serve as a role model to encourage women to train in research. She was considering the possibilities open to her, but when she returned to Columbus and informed her fiancée, Alan, of her job offer at the UNT, he was stunned.

Raquel and Alan's Love Story

Alan told Raquel he loved her very much and did not wish for her to leave him to take a job in Peru. Raquel and Alan had met when they were both residents at Jones Graduate Tower. In the summer of 1973, Alan received his PhD in Mathematics. He left the dorm, but he continued to visit Raquel. The two got along very well from the beginning and they were serious about each other. Raquel met Alan's parents in Rochester, NY, in December 1973. When Raquel was about to receive her Master's in microbiology in August 1974, her parents traveled from Trujillo to Columbus to be at her graduation ceremony. Alan traveled to New York with Raquel to meet them. Raquel was deeply in love with Alan, and she did not wish to cause him distress by leaving him for 2 years. Alan said he had just received word about a postdoctoral fellowship in England the following academic year.

MARRIAGE, FIRST CHILD (1975–1976)

Raquel continued taking microbiology and biochemistry courses through spring 1975. She and Alan were married in Peru that summer. In the fall of 1975, Raquel and Alan traveled to London, England, where Alan had a postdoctoral fellowship in geometry, at the University of London, Westfield College. Raquel had an informal visiting student status in the college's Bacteriology Department. The college was located in Kidderpore Avenue in Hampstead, London. The couple lived in a second story flat at 25 Hamilton Road, on Harrow on the Hill, in the northwest area of London. Raquel and Alan attended Harrow Baptist Church. They were warmly received.

Raquel gave birth to her first son, Kevin, on January 21, 1976, at the Northwick Park Hospital. She received numerous flower bouquets and cards from the Minister and church members. Church members and/or faculty members at Westfield College lent or donated nearly all of the baby's clothing, toys, and equipment. After her son's birth, Raquel threw herself with great zeal into motherhood and home-making. She delighted in cooking and taking the baby out in his English carriage "pram." Alan and Raquel were able to go out at night to London's restaurants and theaters. There was no shortage of babysitting offers from the neighborhood ladies or from the church who were to them much like an extended family. Kevin was beloved by his English "aunties."

RETURNING FROM ENGLAND (1976–1977)

Raquel kept up her membership in the American Society for Microbiology and read journals from the Westfield library. The Fulbright Fellowship carried an obligation to teach at the fellow's home institution for a minimum of 2 years. Raquel planned to fulfill that requirement sometime after returning from England. She had hoped the Fulbright Commission could help her find a position for her husband for fall 1976. However, this was not possible.

Raquel realized that it would be a hardship for her husband to be separated from her and the baby for 2 years if she went to Peru instead of back to the United States. Alan had become assistant professor in mathematics at OSU. Through the efforts of Alan's congressman, Raquel was granted an exception to the 2-year foreign residency requirement and she was able to return to the United States in October 1976. Raquel, Alan, and Kevin settled in a town house in Columbus, near the university. Alan got very busy teaching and researching mathematics at OSU. Raquel decided to be a stay at home mom until summer 1977.

In June 1977, Raquel went back to graduate school to take courses in the Biochemistry Department. Obtaining As in most of her coursework, she was encouraged by her professors to pursue a PhD. She considered a move back to graduate school but, based on her previous unpleasant experiences as a graduate student, she decided to look for a job in a science field instead.

EXPERIENCE AT CHEMICAL ABSTRACTS (1977–1984)

On October 1977, Raquel began to work as an Abstractor in the Biochemistry Department, Toxicology Section, of Chemical Abstracts Service (CAS), a division of the American Chemical Society (ACS). CAS is the world's largest producer and publisher of chemical information and chemical patents from around the world. When she arrived at CAS, Raquel noticed that many of the abstractors were foreign-born former research scientists. She felt instantly at home in that environment.

Her job consisted of reading and analyzing technical journal articles and patents, extracting information, and preparing abstracts and index entries for publication. She was assigned documents that were written in Spanish and Portuguese. She also handled some articles from Italian and French journals. She was glad that her job enabled her to keep up with research developments and with practicing foreign languages she knew.

However, the job demanded 50 or more hours/week. She often took work home for the evening and came to the office on weekends. Moreover, productivity demands were always increasing. Raquel made several suggestions to her management to improve the work flow and to add indexing terms to better highlight danger to the public from toxic substances. Her manager wrote on her evaluation that she was doing a good job but on the negative side, she had "too much social consciousness."

Raquel also joined the local section of the ACS and sought opportunities to be engaged in the leadership of the local section.

In 1980, her second son, Eric, was born. Life was hectic with two children under 5, a demanding job, and active professional life. In 1982, she became Chair of the Committee on Professional Relations. She wrote articles for the Chemical Record, the local ACS newsletter, promoting awareness of the Professional Employment Guidelines, the code of ethics of the ACS, which recommends fair employment practices. She attended the society's monthly meetings and she realized that the speaker was nearly always a man. There were very few women students and no women faculty in the Chemistry Department. However, a few women chemists were hired as staff members and then they were given teaching duties. This practice seemed unjust.

Notable Activities in the Columbus Section of the ACS

Raquel composed a survey based on the ACS Professional Employment Guidelines, the ACS code of ethics, and she proposed to publish it in the newsletter with an article encouraging members to respond anonymously to the questions. She proposed to analyze the data to determine how well the ACS ethics code was known and enforced in chemical companies in the Columbus area. She also presented plans to establish a Women Chemists Committee in the local section that would be dedicated to encourage and recognize women in chemistry. She envisioned that the women chemist's committee would result in greater interest of women to become members and participate in the local section.

There was strong resistance to both these initiatives, especially by the two other women in the 15-member governance board. They argued that no special committee was needed. Women could already get involved in ACS just like they and she did. The survey initiative generated heated discussion when considered. It was tabled without ever coming for a vote. It was eventually dropped from the agenda. The Women Chemists Committee did get the green light to go ahead in the fall of 1984, a little over 2 months after Raquel was fired from Chemical Abstracts.

Getting Fired from CAS

After Raquel became Chair of the Professional Relations Committee of ACS in 1982, she began receiving complaints about managerial misconduct from employees at CAS and at other chemical companies. She listened with empathy and took notes. However, there was little she could do. None of the complainants wanted to do anything about their employment situation, but they wanted Raquel to know. She realized that the employees were afraid. They felt powerless against management, just like Raquel had been as a graduate student. That is why an ethics survey would have been useful.

Research from Spanish-Speaking Countries

Another area of concern for Raquel was the limited coverage of journals from Latin American countries in the Chemical Abstracts publication.Most of the Spanish journals she abstracted were from Spain. It was almost like the 19 countries of Latin America did not exist or did not do any research. Raquel wrote to the CAS Director advocating greater coverage of Latin American journals. She suggested that greater inclusion of Latin American research in the CAS databases could influence the visibility of these countries' scientific output, and this visibility could have a direct effect in those researchers finding funding and collaboration for their research projects. Higher management liked Raquel's ideas, and she was given the green light to research the CAS database going back 15 years. She was to examine the publication of abstracts of all scientific journals from Latin American countries and present her findings at an ACS meeting in Washington, DC. Raquel did a thorough analysis of the databases from 1967 to 1981. She demonstrated that about 67% of the Spanish language research covered came from Spain and 33% from all the 19 Spanish-speaking Latin American countries, combined. The findings were presented at the 185th meeting of the ACS in Washington, DC.

Reporting Misconduct at CAS

Raquel's success in putting together a presentation and publication buoyed her spirits. It also emboldened her to search and find ways to enforce the Code of Ethics (Professional Employment Guidelines) at CAS. On or about January 1984, a new employee, D, was hired to the Biochemistry Department and Raquel was asked to help welcome her to CAS. They had lunch together sometimes and talked to discuss work matters. One day, after work hours, the new hire confided to Raquel that the department's assistant manager had taken her out to lunch that day and had essentially propositioned her. He told her that he liked her and wanted to have a relationship with her. He told her also that in the past several months he had dated another woman in the department, but that the relationship had ended without anybody knowing about it. He promised her he would be very discreet. He indicated that if she agreed to date him, her job assignments would be lighter and her evaluations would be more positive.

Raquel was aghast. The assistant manager in question was married. His wife had hosted a Christmas party for the department at their home. Raquel was dumbfounded. Raquel talked to the other woman mentioned and Raquel confirmed that she had dated the assistant manager for a while, but now he was trying to get her fired. He had written a poor job evaluation of her and had placed her on probation.

In February 1984, Raquel in her role as Professional Relations Chair of the local ACS wrote a letter to the CAS Director informing him, that sexual harassment appeared to be going on in the Biochemistry Department. Immediately, the Director forwarded the letter to the Human Resources Department and they took charge of investigating the matter.

The assistant manager, the new hire, and the woman that was placed on probation were interviewed separately. Within a week, the assistant manager was demoted from manager to abstractor, without any loss in salary. The probation of the other woman victim of sexual harassment was suspended. The new hire soon left for a better job in Philadelphia. On her last day, she told Raquel that she had accepted the job in Philadelphia before telling her of the harassment.

At a meeting of the local ACS Section, Raquel reported what had occurred at CAS and what she had done to enforce the Code of Ethics as was her duty as Professional Relations Chair. A stunned silence followed. Finally someone said, "Are you safe there?" and another said, "No one asked you or expected you to put your job on the line for this committee." Finally, someone said, "pray a lot." Later, other people told her, "you should be looking for another job—sooner or later, they are going to get you."

Soon after the assistant manager was demoted, the department manager began a campaign to find fault with Raquel's work. He began double-checking her work and sent her a dozen of memos pointing out minor mistakes and shortcomings. Raquel was on edge especially because several people had told her, "the manager is out for blood."

Early in the morning on Friday July 13, 1984, the department manager gave Raquel a memo in which he accused her of having "falsified the records" in her time card from the previous week. He indicated he had proof that she had reported in her time about a dozen articles more than she had abstracted. This constituted falsification of records he said, and he asked her to sign a letter of resignation. Raquel was bewildered. Was this retaliation for her uncovering sexual harassment? Were they not glad that the illegal behavior had been discovered and stopped?

The department manager called Raquel throughout the day to ask if she had signed the resignation letter he had left with her. He told her he would give a good recommendation to potential employers if she signed it. Raquel said she could not sign it because she had done nothing wrong deliberately. In a panic, she had called the CAS director to inform him of this threat to her job but the director told her, "it is out of my hands—good luck."

Taking Legal Action against CAS

After being terminated, Raquel applied for unemployment compensation and it was granted. Chemical Abstracts appealed the decision on the grounds of termination for cause. Raquel requested a hearing in person. Raquel presented evidence of good to excellent job evaluations for 7 years prior to her reporting sexual harassment. At the hearing, the department manager testified that she had been terminated because she had reported incorrectly the number of abstracts she had prepared in a particular week. When questioned about whether he had physically counted the articles in question to establish a discrepancy between the number of articles reported and the numbers of abstracts produced, the manager indicated yes.

When asked if he had been on a mission to find fault with Raquel's work ever since she had reported sexual harassment, he responded no. When asked if he had checked the production records of each of the 95 abstractors in the Biochemistry Department to see if their time cards were filed accurately, he replied yes.

When asked how many hours had it taken him to pull out the time cards, go over the production records of 95 employees each of whom had processed about 100 articles and to physically count the articles to verify the correctness of all time cards, he was silent. When the question was repeated to him, he admitted that he had not physically counted any documents nor verified anybody's production records.

The Unemployment Compensation Board dismissed Chemical Abstracts appeal stating "Chemical Abstracts alleged that the claimant had falsified production records. At a hearing, the weight of the evidence did not substantiate said allegations."

Raquel filed a claim against CAS with the Ohio Civil Rights Commission and later filed a civil suit for wrongful termination in violation of Title VII of the Civil Rights Act of 1964. Raquel's lawyer, Dennis Pergram, accepted the case on a contingency basis. CAS retained, Bricker & Eckler, the largest law firm in town. The motions the law firm put forth were voluminous and daunting. But, in the end the simple truth of the matter prevailed. Raquel and her attorney accepted a cash settlement offer from CAS in 1988.

RETURN TO GRADUATE SCHOOL (1984–1987)

After leaving Chemical Abstracts in 1984, Raquel returned to graduate school at OSU. She applied to the Graduate Business School to the Master's in Labor & Human Resources Management program. She wanted to understand the employer–employee relationship in the context of a business infrastructure. She felt that improvements could be made at the workplace if ethical principles could be codified into optimal employment practices.

In the Graduate Business School, Raquel found a very supportive and welcoming atmosphere. To start with, most of the classes were held in the evening to accommodate older students who had day jobs. Their instructional method was andragogic as opposed to pedagogic. Professors presented topics and provided insights, but students were expected to contribute their own perspectives.

Raquel read voraciously every book she could find on management theories past and present, motivation theory, compensation theory, and so on Raquel took a special interest in Employment Law, which was part of the curriculum and she became very familiar with the Employment-at-Will doctrine and Title VII of the Civil Rights Act of 1964.

Raquel excelled in her classes and made many friends in business school. She organized several local and national conferences in the Professional Relations Division of the ACS where she invited her professors to speak to chemists about tenets of Employment Law and other aspects of the business infrastructure that affected the employment of chemists. These activities contributed to Raquel's becoming very well known at the local and national level within the chemistry community. She was nominated numerous times as a candidate for office and was elected to national offices of increasing prestige and responsibility. Raquel was appointed to the national Women Chemists Committee for 1988−1990. She became national Chair of the Division of Professional Relations in 1992 and served in that office from 1993 to 1994. She was also an elected national Councilor for the Association for Women in Science (1996−1998).

Family Additions, Changes

Leaving CAS had a silver lining. Raquel went back to school and became pregnant for a third time. This period was happy with the arrival of her daughter, Susan, in July 1985. She often said that had she stayed in CAS, she would not have had another child because of her already too full plate. The care of baby Susan was divided more equitably between Raquel and her husband Alan. Both were in their 30s and they had become parents again and graduate students again.

Alan was studying for a second PhD, in computer science, at OSU from 1982 to 1988. In fall 1988, Alan took a job as an associate professor of computer science at the University of Alabama at Birmingham (UAB).

Raquel had been an Intern at Battelle Memorial Institute, Project Management division for 6 months in 1986−1987. Early in 1988, Raquel was offered a job with the state that involved a lot of travel. In the summer of 1988, Raquel was also offered an administrative job at UAB. For a while, she considered taking the job at UAB and moving to Alabama. But in the end, both Alan and Raquel agreed that their kids, aged 12, 8, and 3 would have better education opportunities in Columbus. Raquel would be better able to balance taking care of the kids and attend to her own intellectual development if she could work as a consultant.

ESTABLISHES OWN COMPANY (1988–PRESENT)

Raquel became convinced that a lot more work was needed to flatten organizations and to bring about a humanistic work environment where the contributions of each person were valued and their talents used to the fullest extent possible. She realized that there was too much fear in the workplace. In a fearful atmosphere, wrongdoing could go undetected and unreported for a long time to the detriment of everyone. Research was needed into best management practices at home and abroad.

She incorporated a company called Technical Support Incorporated dedicated to management consulting, organizing conferences, and supporting organizations to promote compliance with employment laws and the application of ethical principles in the workplace technical recruitment and communication strategies including culturally conscious translations were also offered.

NATIONAL CONFERENCE ON DRUG TESTING (1989)

Raquel organized and chaired a panel on "Drug Testing in the Workplace: Chemical, Legal, and Ethical Implications" for the national ACS meeting in Denver, CO, in 1989. The speakers were eminent chemistry professors and a lawyer who had just testified to Congress on the subject of drug testing. In those days, employers were rushing to institute preemployment drug testing, but the chemistry involved in the testing methods was flawed and results could be misleading. The legal and ethical implications of testing were very significant. The conference was packed and generated a buzz. Reporters from the *Denver Post*, *Philadelphia Inquirer*, and *Chicago Tribune* covered the event and interviewed the panelists. Raquel was quoted in the *Denver Post* and in *Chemical & Engineering News*.

NOBEL LAUREATE SPEAKS AT FIRST CONFERENCE ON ETHICS IN SCIENCE (1991)

In 1990, Raquel began organizing and planning for a conference series entitled "Ethics in Science, Technology, and Medicine." The first such conference took place at OSU in April, 1991. Roald Hoffmann, Professor of Chemistry at Cornell University and Nobel Laureate in Chemistry, spoke eloquently about Chemistry, Democracy, and Social Responsibility of Chemists to a standing-room only audience of 150 people. Raquel introduced the speakers and four panelists representing fields of expertise in philosophy, economics, law, and theology.

The conference was a great success and earned respect for Raquel's skill as organizer of well thought out conferences on timeless topics. The conference was cosponsored by the Association for Women in Science and the ACS. The conference was recorded, and because of demand, videotapes were made of the conference and at least 30 tapes were sold to interested parties. It was the first time that a conference had generated revenue for the societies.

After this event, Raquel was nominated and selected for the Outstanding Woman in Science Award and received that prestigious award from the Association for Women in Science of Central Ohio in 1992. Her chemistry friends nominated her for Chair Elect of the National Division of Professional Relations and she was elected to that post in 1992.

Raquel went on to organize seven Ethics in Science and Medicine conferences until 2002.

TEACHING AT THE COLLEGE OF MEDICINE (1991–2008)

After Raquel published an article about the importance of culturally conscious translation for the local community newspaper, medical students approached her to give them a short course on Medical Spanish. They agreed on the terms and she taught a 5-week course entitled Survival Medical Spanish. The students liked the course and wished to have more. They petitioned their dean and in 1992, Raquel was named a Mini-Module Instructor on "Medical Communication with Spanish-Speaking Patients" a 5-week, 20-contact-hour course. The course, which was limited to 12 students, was very popular.

Raquel introduced a variety of artistic elements into her teaching, including ceramic figurines from ancient pre-Inca cultures depicting health and healing concepts.

ESTABLISHING MEDICAL STUDENT EXCHANGES (1993)

In 1993, she established a bilateral medical student exchange program between the OSU medical school and the UNT. Students from Peru were seldom able to come to Columbus, but in 1998, eight students from Trujillo came. The exchange program resulted in great opportunities for medical practice and unforgettable experiences for the students who stayed at the homes of Peruvian families. When students came back to Columbus, they made wonderful comments such as this:

> My home-stay was one of the most wonderful experiences of my life. The Burga family not only took me into their home, but into their hearts as well. I became a member of the family, and was very saddened to leave them when the time came. We still keep in touch over email and the occasional phone call, and I look very much forward to the day when I can visit them again. *Kimberly B. (2005)*

FOUNDING DIRECTOR OF LA CLINICA LATINA (2000)

In 1999, Raquel founded a community group, the Latino Health Alliance and in 2000, she submitted a grant application to establish a free medical clinic for uninsured Latinos. She started La Clinica Latina on December 19, 2000, with a $10,000 grant from the Ohio State University Medical Center. The clinic, which continues to this date, offered 720 patient visits in its first year. The clinic has also served as a service-learning site for students taking courses to improve communication with Latino Patients.

Raquel has given numerous lectures on "Cultural Adaptation in Medicine" and "Medical Communication with Latino Patients." In 2003, she received the Health Hero Award from the University of Illinois at Chicago. In 2008, she received a grant from the Area Health Education Center to produce a video entitled "Communicating with Your Latino Patient."

DIRECTING WOMEN IN SCIENCE DAY PROGRAM (1999–PRESENT)

The Women in Science Day is a 25-year-old tradition at the OSU. This program brings 300 gifted girls to the OSU campus for a day of science workshops led by women scientists. It has been under Raquel's leadership since 1999 when she received a $50,000 grant from the Battelle Endowment for Technology & Human Affairs. She restarted the program that had been discontinued for a while. The program's participants include bright girls in Grade 6 to Grade 8 from around the state. The Women in Science Day is highly regarded by participants and draws a number of visitors. In 2009, the program was chosen as a featured program for the "Doing Something Great" series, and it was one of the programming highlights for 2009 (see http://www.osu.edu/features/2009/womeninscience/).

The Women in Science Day Program has been recognized externally by the Ohio Women's Policy & Research Commission (2001) and internally by the Multicultural Center (2008). The program at OSU is the oldest, largest, and best-known program of its kind nationally. A quick Google search brings up the OSU program posted at http://awisco.osu.edu/. The next highest ranked programs in Google are at the University of Wisconsin, Stevens Point, and at Columbia University.

The Women in Science Day Program includes an essay and picture contest entitled, "Honoring Women in Science through History," which draws on the imagination and sensibilities of women scientists as viewed by the eyes and minds of middle school girls.

ETHICS BOWL CASE WRITER, JUDGE, AND ACTIVE MEMBER ASSOCIATION FOR PRACTICAL & PROFESSIONAL ETHICS (1997–PRESENT)

Raquel's involvement in the work of Association for Practical & Professional Ethics comprises a wide range of interdisciplinary efforts. She presents often at the annual meeting, chairs sessions, and is a founding member of the Diversity Committee. Since 1997, she is a member of the Case Committee, which writes the cases for the Regional Ethics Bowl in which over 100 colleges and universities participate. She is particularly interested on codes for ethical intercultural communication that acknowledges and respects different world views, particularly in health and healing. She believes that many persistent, unresolved ethical issues are gender, color, and race-related.

There are vast opportunities for those who wish to expand mankind's body of knowledge by organizing and participating in conferences within academia without necessarily having a tenure track position. There are many areas of human endeavor that merit closer examination by minds trained scientifically and humanistically. One such area is the field of ethics. Ethical principles such as beneficence, non-maleficence, autonomy, distributive justice, and others should be a part of basic training. These principles are not just for philosophers or scientists, but they are also for students in all fields. Ethical principles should inform the reasoning of every educated man or woman regarding the practice of their profession.

SIX LESSONS LEARNED

1. Importance of principles, preparedness, and poise in public speaking.
   One must read, be knowledgeable of moral principles, and be prepared to defend them. Knowledge and preparedness enables us to speak without fear and with confidence, authority, and credibility.
2. Community groups and professional societies can be leadership training grounds.
   Student organizations and professional societies have been Raquel's leadership training ground. In an era of primarily electronic communication, the human exchange in face-to-face communication must be preserved.

3. Instigate partnerships between professional societies and academic institutions to increase impact and reduce perils of advocacy for ethics, gender equity, and health justice.
4. Mentors may not exist. Beware of the false mentor who will try to manipulate the junior person for self-gain. Some women have "queen bee syndrome" and they are not helpful. It is possible to find sympathetic allies among white males.
5. Ignore slights. Successes and failures are twins. Take advantage of your failures.

    Always keep in mind the larger perspective, not just your short-term survival.
6. Be present and involved with your children. Be a good parent and a good partner. Your spouse and children are your best friends, partners, and allies. You must cherish them.

6

Creating My Own Path

Dixie L. Dennis

Austin Peay State University, Clarksville, TN

> *Each person's life*
> *represents a road*
> *toward himself [sic]*
> **(Herman Hesse, 1991)**

BRIEF BACKGROUND

In general, to understand the place at which a person finally arrives sometime later in life, it is helpful to know from whence this person came. Briefly and personally, I came from a small town (population 1,000—then as well as now) populated with people who had very traditional religious values and parents who "did the best they could do," which is a statement people typically make to indicate that things weren't particularly right. To explain, in general, my father was verbally and emotionally abusive, and my mother was distant as she, along with my sister and me, tried to avoid his wrath and judgment. Both explicitly and implicitly from my religious mother and the southern town's people, I learned "Be content with what you have"; "Women are submissive to men"; "A woman's place is in the home"; and "You made your bed, now lay [sic] in it." And, as a controlled, belittled, and shame-filled person, I "took up my cross" for about the first four decades of my life. Then, as my oldest son opened the front door, leaving the house to enroll in college, he turned back to me and said/asked, "Mama, you look so depressed; why don't you take a course or two in college?" Not wanting to let him down, I did it—I took the biggest, and most nontraditional, step for someone who grew up as I! Possibly, it is true that "when the student is ready, the teacher will come." But, why had it taken me so long to be ready for that teacher? Rainier Maria Rilke (1934) wrote:

> Be patient toward all that is unsolved
> in your heart...
> Try to love the questions themselves...
> Do not now seek the answers,
> which cannot be given
> because you would not be able
> to live them.
>> And the point is, to live everything.
>> Live the questions now.
>> Perhaps you will then gradually
>>> without noticing it
>>> live along some distant day
>>> into the answers.

THE EDUCATIONAL PROCESS

Particularly important in what I will describe about pursuing my education is that I never had a goal to "be" anything or "become" anybody. I never thought at the outset about a particular job. I merely was trying to find something positive so my spirit as a person would not be killed. In hindsight, I realize that I was meant to be something and become somebody. The lesson for me was that, even for folks who are no longer spring chickens, it's never too late to start becoming who you are meant to be.

After obtaining a bachelor's degree in chemistry (1989; GPA 3.9/4.0) and a master's degree in public and community health (1996; GPA 4.0/4.0), I earned a doctorate in health education (2000; GPA 4.0/4.0). I do not have the words to explain how fearful and intimated I was with each educational start as well as throughout each step of each journey. I surmised that, all along the way, everyone would be younger and probably smarter. Although that was a fairly accurate account, ☺ I started—and I finished. Along the way, there were incidences of prejudices—not just about gender but about age and intellect.

While working on my bachelor's degree, one particular incident is exemplar of prejudice related to age and intellect. One much-younger female student who repeatedly said, almost in a sing-song manner, "I'm a math major," trying to let students in chemistry, with me, an older student in particular, know she was smart and likely smarter than anyone else. She sat immediately in front of me in the Physical Chemistry course, the most dreaded, and believed most difficult, undergraduate course on any campus. After getting the results from our first test, a test in which I was sure I'd failed and a test in which I'm sure she believed she had earned the top score, she turned to me and asked, "What did YOU get?" I handed her my paper wherein it was revealed that I had received a score of 94. She said, as if in disbelief, "YOU made better than I did!" From then on, she lessened the sing-song mantra, and, thereafter, she frequently asked me for help to understand the difficult material in that course. The lesson in this scenario is that positive pay-offs exist for overcoming fear and intimidation of others, including those who perceive they are "better than"—in any sort of way.

One of my greatest life experiences (i.e., obtaining my doctorate) came from a personal tragedy. In short, I finally decided to leave my alcoholic husband. As I considered how to support myself, a professional friend suggested that I get a doctorate. Me, get a PhD? I had never considered it, but, before long, I submitted an application and was accepted into the program. Important to this situation is that without the chaos and craziness at home, which was similar to, but worse than, that during my upbringing, I would never have gotten "fed up" enough to want to live more positively. Like the words in a Gladys Knight song (1973), an alcoholic husband was, at that time, "the best thing that ever happened to me," because I found me. Since that time, I have remarried. I married a wonderful, loving man who has encouraged me to be everything God meant for me to be. He's the REAL "best thing that ever happened to me." Due to a 6-year battle with cancer, he died 2 years ago. I'm glad I had him in my life to teach me that I am a good and valuable person.

I had many other lessons to learn as I kept becoming an even better new-found me. One lesson that I still frequently think about is what happened in the second year of working on my doctorate. I realized that I was a fairly good writer. I decided that I would write and submit an article to the teaching ideas column in a top health education peer-reviewed journal. I must admit that writing and publishing were pretty far-fetched ideas given that, up until that time, my writing endeavors had been confined to class and lab papers (although, I worked hours at a time to perfect each one), with the majority of my writing delimited to making grocery lists. When I excitedly told a professor, whom I respected, that I was thinking about submitting a teaching idea to a journal, I was told in no uncertain terms, "When you are ready to publish, I'll let you know." What I didn't tell the professor was that I already had submitted it. That old feeling of shame overtook me. I had not even considered that I wasn't ready—or that I'd be rejected. What if I wasn't ready? What if I never would be? Then, something important happened. That article was accepted—and without revision. Moreover, 8 years after this event, I was asked to take over the editorship of that teaching ideas column! The reflective lesson is that, if you hear negative voices—either real or those in your head—you don't have to listen to them. Rewards exist for rising above negativity. Another lesson a person could glean from this story is that, although it is ideal to be encouraged and supported, you can do positive things without either.

After the publication of that article, I was bitten by the research and writing bug. In about 9 years, I have been successful in book publishing with:

three book chapters (Definition of death, funerals, and near-death experiences, in M. Brennan, 2014),
a textbook (*Living, Dying, Grieving*, 2009), and
a book chapter (Biological basis of spirituality, in Speck and Hoppe, 2007).

Moreover, to date, I have had 27 peer-reviewed journal articles published in various journals, and I have two research projects in progress. Also, I have had 18 invited articles published. Through these experiences, I have worked with people across the United States and in various academic disciplines. Many times, I was first author, but, at other times, I was second or third author. If I had listened to that once-respected professor, and pulled the article from consideration, I likely would have missed meeting and working with so many wonderful people who enriched my life. In the final analogy, my writing contributed to my rise in academic, and it became something more. As Charles Schulz, the creator of the famous comic strip, "Peanuts," was reported to have remarked when asked why he drew cartoons, "Why do musicians compose symphonies and poets write poems? They do it because life wouldn't have any meaning for them if they didn't. That's why I draw cartoons. It's my life" (Lickteig, 2000). Like Mr. Schulz, writing had become my life. Good grief, as Charlie Brown might say!

Another outcome of my passion to write and publish is that I was asked to teach a research methods course, which led me to the position of research coordinator for two doctoral programs, which, eventually, led important others to believe in my value as an administrator. To jump ahead a bit, on each campus I've worked, I've been thought of as THE editor. At the outset, though, I did not want to teach that first assigned research methods course, and I worked long hours to understand concepts so that I could present them to appear knowledgeable.

Not too much time passed before I was an expert in teaching research methods to undergraduate and Master's level students. A few of those students told someone in administration from a close-by university that I was a good, and fair, teacher of research methods. A dean at that university called and asked if I wanted to teach research methods to doctoral students and become the Research Coordinator for two doctoral programs. Then, at a different university, because of this leadership experience, I became chair of a department. Soon, I was dean and director of grants and sponsored research. Three years ago, I became the Associate Provost for Grants and Sponsored Programs and Dean of the College of Graduate Studies. I have learned that success looms for many people who, like what is written on the tag of the "Life is Good" tee shirts, "Do what you love, love what you do."

Another example of prejudice in an educational setting happened to me, a "southerner," while I was obtaining a graduate degree from a "northern" institution—during, of all courses, a psychology class. The professor of the psychology course often smiled when I answered a question he asked of me. I must admit, though, that I sounded very different from the rest of the students in the course who hailed from around the area. Eventually, I knew he was making fun of me and, without going into detail, it was evident he believed I wasn't very bright. I took an opportunity to save my bruised ego. On the next test following my realization (The tests in that course were standardized and, therefore, extremely difficult.), I made 100%. No one else did— on that test or any other in that course. As you may realize by now, my way of surviving is to show myself that I am not what someone's label suggested. A person should feel good about being herself. As Dr. Suess, American writer and cartoonist of the 1900s (Geisel, 2009) once said:

Today you are You,
that is truer than true.
There is no one alive
who is youer than you.

WORK IN ADMINISTRATION

For a bit of background, in my first professorship position in 2000, I was greeted by a man, who would become one of my dearest friends, with "I want you to know that if you had applied for this position 40 years ago, we would not have hired you; we didn't hire women." I believe he still longed for the 1960s. I understood from where he came, though, because I came from that same place. In my next two teaching jobs, the you're -professionally-flawed-because-you're-a-women was less visible. I had gained many valuable skills by then and was increasingly regarded as a good scholar, teacher, writer, and leader. In this time period, I allowed myself a possible rejection event when I submitted applications for service on national boards. To this day, I cannot believe enough people knew me, or believed in me, to vote for me. I took another chance and was elected to both boards. As Dr. Brent Dennis wrote in 1990:

> Life is not a game of chance.
> Life is a journey of choices.
> Negative and positive,
> The choices,
> And the consequences,
> Are ours.

I served as chair of the National Commission for Health Education Credentialing, Inc., Board of Commissioners and member of the American Association for Health Education Board of Directors. Upon reflection, I realize that, after experiencing success, other rejection-risky steps are easier to take. Regardless of ease of risking, though, I am convinced that, given I had no training about being a successful leader, my successes in academic and health education professional pursuits were due to the following actions. I:

> wear a suit (i.e., I dress professionally).
> work hard. (My typical day goes similar to yesterday:
> Answer e-mails [100–200, many of which require thoughtful answers]
> Take phone calls
> Sign papers
> Make decisions from people who report to me
> Review 1–3 journal submissions and assign the manuscripts to reviewers
> Engage in a national conference call
> Review a university policy regarding compliance issues
> Attend a faculty roundtable discussion
> Work on a federal year-end report of a $2,000,000 grant
> Review a federal grant
> Finalize more plans for a community/university event (expecting 4,000 in attendance)
> Attend a meeting with the provost).
> smile (Why not?).
> make it a practice to be kind, courteous, punctual, organized, and conscientious.

To me, living this type of day and being this type of person is just doing my job. I surmise that many people just don't do their jobs and those of us who do look like super stars. I wonder. In the end, these traits are requirements for anyone wanting to make administrative advancements in academic—even for women who have not experienced ageism (older people have little, or nothing, worthwhile in leadership), sexism (men are natural leaders; women are not), or cultural elitism (southern women are dumb). It may be difficult for some readers to realize that, in times past, three appropriate words were the weatherman, the postman, and the chairman. Now, mind you, I NEVER want to get a job because of being a woman. I want the job because I'm the best applicant. Today, I get that opportunity; in the 1960s, I did not have the opportunity to be the weather person, the mail delivery person, or the chair person. For those of you who did not live the leadership *status quo* of the 1960s, know that you are standing on shoulders of people who lived it. And, to some extent, we all still recovering. For example, when walking into a room full of academic administrators, a person is likely to see many more men than women. For sure, women still are recruited less often to academic leadership positions than men, and they make lower salaries and are less recognized through awards (Dominici et al., 2009). Apparently, women have a long way to go, but, "the times, they are 'a changing," and that's the important part.

> The line it is drawn
> The curse it is cast
> The slow one now
> Will later be fast
> As the present now
> Will later be past
> The order is
> Rapidly fadin'.
> And the first one now
> Will later be last For the times they are a-changin.'
>
> *(Dylan, 1963)*

My role in administration began in 2005 when I found a job announcement for department chair in a geographic area close to my children. I got the position and, therein, as previously mentioned, began my stint as an "official" academic administrator. When I started as chair, it was obvious that the faculty in the department were angry, depressed, and suppressed, and they were not interested in conducting research, a practice that is important to all faculty, including those in a teaching university who want to be better teachers. In addition to these issues, the person whose place I took, told me, "You should run this place more like a man. If not, they're going to see you as weak." Interestingly, I got through that encounter without expressing the expletives I silently formed in my mind. During the weekend that followed, I spent hours looking for articles about women leaders and found just what I needed, that is, women typically make better leaders than men. That, of course, is not always true, but it offered me the chance to stand up for women. On the following Monday, after my colleague's proclamation about male superiority, I placed a "clarifying" article in his department mail box. He never again told me how to lead.

While I was chair in that department, I was determined to provide an atmosphere in which faculty could flourish. Today, they thank me for what I did, but, to be truthful, all I did was be nice to them, brag on them to others across campus, give them a say in when/what they taught/did, and I showed them, and helped them, with doable research opportunities. People can get themselves out of darkness when someone shines a light on the door. That's what my son had done for me; that's what I did for them. In my position as departmental chair, in addition to helping faculty become their potential, I turned around requested work to the dean quickly. Soon, an opportunity arose. This influential dean asked me to be his associate dean. With his notice of my leadership qualities, I was on a path where other opportunities soon came. Musil (2009) wrote, "Being championed by an influential person matters significantly as does having leadership credentials" (para 8).

To date, I have been interim dean of two different colleges in my university. In one, I did not have a terminal degree in a discipline in the college. But, "people said" I was a born leader. Upon reflection, I wondered if they were right, but I knew I had many barriers to cross before I became what I may have been born to do. As I consider my life, I frequently wonder if my life problems were serendipitously "just there" and that I eventually made good choices about how to solve them—or I wonder if Albert Einstein was right, when, in an autobiographical hand-written note, he wrote, "Something deeply hidden must have behind things" (Bartlett, 1980), meaning that, all along, I had the "right" problems placed in my life to overcome and eventually become who I am today. I guess I'll never know if my problems were there by accident or design, but I like the sound of the latter better. Nevertheless, a good lesson to learn from my life is that people should be thankful for problems, because they offer chances to grow.

Throughout my rise in education and administration, specifically, and life, in general, I do not believe people were against me. They were just reacting to me from where they were in life, which was filled with all of their personal past and present life experiences. In every good and bad event in my life, both male and female were perpetrators of prejudice. And, both males and females helped me along my life and leadership journeys. In other words, failure is not because of what someone does to someone else, it is the positive stride the victim makes in spite of what was said or done to him/her. It is important not to let any form of prejudice limit what a person can be and become.

In 1994, Bob Seger wrote about a lesson that can be applied to overcoming problems in both our professional and personal lives. He wrote and sang,

> I can sit here, in the back half of my life
> And wonder when the other shoe will fall
> Or I can stand up, point myself home
> And see if I've learned anything at all…
> Time to lock and load
> Time to get control
> Time to search the soul
> And start again.

In an Al-Anon book, *Courage to Change* (1992), another person explained overcoming problems this way:

> Though no one can go back and
> Make a brand new start,
> Anyone can start from now and
> Make a brand new end.

In summary, I hope the lessons I learned in my unplanned climb in academia help you live the productive and meaningful personal and professional life that you were meant to live. To get to that place, I offer my best wishes on your road to find yourself as you live into each new:

> start;
> end;
> problem;
> lesson,
> question, and;
> answer.

References

Bartlett, J., 1980. Familiar Quotations. fifteenth ed. Little, Brown and Company, Boston, MA.

Courage to Change: One Day at a Time in Al-Anon II, 1992. Al-Anon Family Group Headquarters, first ed., Virginia Beach, VA.

Dennis, B., 1990. Responsible problem solving: using perceptual control theory. J. Juv. Justice Deten. Serv. 12 (1), 11–17.

Dennis, D.L., 2007. Biological basis of spirituality. In: Speck, B.W., Hoppe, S.L. (Eds.), Searching for Spirituality in Higher Education. Peter Lang Publishing, New York, NY.

Dennis, D.L., 2009. Living, Dying, Grieving. Jones & Barlett, Sudbury, MA.

Dennis, D.L., 2014. Definition of death, funerals, near-death experiences. In: Brennan, M. (Ed.), The A-Z of Death and Dying: Social, Medical, and Cultural Aspects. Greenwood, Santa Barbara, CA.

Dominici, F., Fried, L., Zeger, S., 2009. So Few Women Leaders: It's no longer a pipeline problem, so what are the root causes? Academe [online]. Available at: <http://www.aaup.org/AAUP/pubsres/academe/2009/JA/Feat/domi.htm>.

Dylan, B., 1963. The times they are 'a changing, Sony Music Entertainment.

Geisel, T., 2009. Be who you are [online]. Available at: <http://thinkexist.com/quotation/be_who_you_are_and_say_what_you_feel_become/341011.html> (accessed May 31, 2014).

Hesse, H., 1991. Courage to Change. first ed. Al-Anon Family Groups, Virginia Beach, VA.

Knight, G., The Pips, 1973. The Best Thing That Ever Happened to Me. On Imagination [record.]. CA: MGM.

Lickteig, M., 2000. Cartoonish Charles Schulz dies on eve of last comic strip [online]. Available at: <http://www.boston.com/news/daily/13/shultz2.htm> (accessed July 13, 2010).

Musil, C., 2009. OCWW I Vol 35, Issue 2 I Director. [online] Aacu.org. Available at: <http://www.aacu.org/ocww/volume35_2/director.cfm> (accessed May 31, 2014).

Rilke, M., 1934. Letters to a Young Poet. first ed. W. W. Norton, New York, NY.

Seger, B., 1994. On Lock and Load. Woodland Digital Studies, Nashville, TN [CD].

7

Pathways in Athletic Administration

Lenora Armstrong

Department of Health, Physical Education & Exercise Science,
Norfolk State University, Norfolk, VA

PART ONE: THE ACADEMY

This essay explores the career pathways for becoming an athletic director (AD) at National Collegiate Athletic Association (NCAA) Divisions I, II, and III institutions based on gender, race, and ethnicity. There is a lack of diversity in athletic administrative positions, especially at the AD level, within colleges and universities. However, there is limited scholarship on how gender and race and ethnicity may affect career paths up to the ADs and especially in NCAA Divisions I, II, and III.

While women hold a healthy number of the total jobs in collegiate athletic administration, they are not well represented in the senior positions. Coakley (2001 and 2009) argued that jobs for women in coaching and administration are limited because men control most sports programs; although women's sport programs have increased in number and importance, these teams are still in less powerful positions within the sports hierarchy than those of men, which connect to the underrepresentation at the highest levels of power in sports. Finally, Title IX, which opened the doors for women as participants in sports, seems to have led to closed doors for women in top collegiate sports administration jobs (Abney and Richey, 1992).

Statistics show the low numbers of minority women in AD roles ranging from 0.6% for Blacks and 0.4% for other minorities in 1995–1996 to 1.3% for Blacks and 0% for minorities in 2005–2006. In the various categories by divisions, the numbers range from 0% to the largest percent of 2.5% at a Division II HBCU (NCAA, 2007–2008).

Navigating Academia: A Guide for Women and Minority STEM Faculty.
DOI: http://dx.doi.org/10.1016/B978-0-12-801984-9.00007-9.

79

Although minority women have made strides in intercollegiate athletics, few have achieved positions in athletic administration in general, or in AD positions. Most African American women are concentrated in lower level positions such as secretary, graduate assistant, assistant coach, administrative assistant, assistant AD, compliance officers, life skills coordinators, or athletic academic advisors/coordinators (Abney, 2000; Wicker, 2008). Opportunities for minority women have increased in the positions of graduate assistant, academic advisor, senior woman administrator, and intern (NCAA, 2007–2008).

In order for women to reach equality with men in coaching and athletic administration, women will have to both regain lost influence in women's athletics and begin to enter positions of authority in men's athletics (Feminist Majority Foundation, 1995). According to NCAA (2008–2009), the initial 1989 report on barriers indicated that, in general, women in athletics administration and coaching were content with their careers. However, some respondents described mixed emotions about their career in intercollegiate athletics. Some described the perception of women as "second-class" citizens. Respondents also indicated women's involvement with sports was often perceived as an association with lesbian and/or masculine stereotypes. Additional frustrations highlighted within the report revolved around athletics politics and general working conditions. Female administrators and coaches in 1989 attributed the lack of female interest in intercollegiate athletics careers to a perceived interference with marriage and family duties as a result of time demands. Other factors highlighted in the reports were a lack of initiative for involvement in athletics, stress, lack of advancement and opportunity, and low pay.

The current NCAA (2008–2009) findings indicate that female administrators would still be an intercollegiate athletics administrator if they were to start over again and agree that they encourage current student athletes to consider intercollegiate athletics as a career. While the majority of female administrators indicate satisfaction with their current overall employment, some indicated dissatisfaction with the gender equality within athletics departments and the equality of race/ethnicity in athletics departments. They also feel there are qualified women who do not apply for intercollegiate athletics administrator positions because of time requirements and may leave their careers in athletics administration because of family considerations.

Still many women who work in sport organizations face the burden of dealing with an organizational culture that they have had little or no role in shaping, which contributes to high turnover among women (Coakley, 2001, 2009). Barriers to career opportunities for women in sports are being hurdled and knocked over, but the forces that have limited opportunities in the past still exist (Coakley, 2001). Even though the opportunities for women in the sport profession may continue to shift toward equity, many people resist making the structural and ideological changes that would produce full equity (Coakley, 2009). Those who hire leaders often unconsciously let pervasive sexist stereotypes of women (such as lack of assertiveness and knowledge) influence their decisions (Delano, 1990). Theberge (1983) and Dunning (1986) note, sport has traditionally been seen as a place where sexist versions of masculinity are constructed and maintained, thus sport is not perceived as proper place for women.

Sexism is structured in major societal institutions such as the family, economy, education, politics, media, and sport (Delano, 1990). Other oppressions that influence the lack of women in athletic leadership positions are diversity among women and their particular identities, situation, or context meaning racism, classism, and heterosexism (Delano, 1990, 1988). For example, homophobia affects the hiring of female athletic leaders because it affects the decisions of those who feel that lesbian athletic leaders do not present a proper public image or cannot serve as proper role models for girls (Delano, 1990). In another study of Delano (1988), racism limited Black females' employment to an economically poor urban area where athletic administration jobs did not offer monetary compensation, release time, or status benefits. Thus, the Black women interviewed who had both economic and time constraints due to family responsibilities did not find such jobs attractive (Delano, 1990).

Career Pathway Research: In General

The general scholarship on career pathways is weak in treatment of non-White, male administrators. The foundational studies on managerial career pathways (e.g., Holland, 1959; Parson, 1909; Super, 1953) were conducted during an era of a mostly White, male-dominated work force (Luzzo, 1996). Women, minorities, and the economically disadvantaged have been methodically omitted from career pathway research due in part to their low numbers in managerial positions (and especially senior positions) in US businesses (Leong, 1995; Parham and Austin, 1984; Sue and Sue, 1990). Criticism of the lack of diversity in career pathway research has pointed to the fact that women and minorities tend to be in nontraditional careers that are not typically captured in career pathway studies; the gender and racial biases, stereotypes, and expectations about how women and minorities are like and should behave that make capturing their experiences in career pathway studies difficult; the underappreciation of the multiple roles and jobs that women and minorities often have in the workforce and the relationship of these to undercounting in career pathway studies; and the low numbers of administrators in these groups (particularly at senior levels) that make statistical analysis difficult in career pathway studies (Abney, 1988; Deller, 1993; Heilman, 2001; Lough, 2004).

Perhaps the lack of role models and mentors has been a deterrent (Abney, 2000). According to Smith (1991), planned interventions with adult role models and mentors do appear to have a positive effect on the career behavior and aspirations of African American youth. In most traditionally White institutions, the African American women athlete lacks African American women administrators and coaches, with whom she can identify (Abney, 2000). In most traditionally Black institutions, African American males occupy a large percentage of the positions in sport (Abney, 2000). At some point in time, the glass ceiling presents an impenetrable barrier in the managerial careers of women and people of color, and career pathway research has not, in general, found ways to surmount this issue (Morrison et al., 1987). In review, although the general research on managerial career pathways in the United States provide a foundation for sector career pathway studies, the general studies have not done a good job of looking at diversity issues in managerial career pathways.

Within collegiate athletics, there is ample scholarship on issues such as Title IX and its impact; the preparation and training of athletes; the status, barriers, and underrepresentation of women and minorities in the AD role; and race and ethnicity issues in the NCAA Divisions I, II, and III. There is limited scholarship on the low percentages of women and minorities in athletic administration, their career development, and advancement (Wicker, 2008). Beyond collegiate athletics, in the larger field of sports management, there are few studies that reveal career patterns experienced by sport management personnel in the sport marketplace (Kjeldsen, 1990). Within collegiate athletic administration, there is a paucity of studies on career pathways, specifically relating to women and people of color.

Conventional wisdom holds that careers of the ADs evolve in a common and sequential manner. This pattern comprises five steps: college athlete, high school coach, college coach (head or assistant), assistant or associate AD, and college AD (Fitzgerald et al., 1994). At the time of the Fitzgerald et al. study, virtually all athletic administrators were White males. There has been a great deal of literature on career pathways between the early 1970s through the 1990s (e.g., Gerou, 1977; Bond, 1983; Acosta and Carpenter, 1985; Danylchuk et al., 1996) and then the literature became limited after 2000 (e.g., Coakley, 2004; Cohen, 2001; Grappendorf and Lough, 2006; Lapchick, 2008; Tiell, 2005). There has been little research on this career pattern in the decades since the Fitzgerald et al. (1994) study. It is unclear whether background characteristics such as race and gender would be expected to impact the advancement after an individual enters an occupation (Spilerman, 1977).

Although the door to athletic administration opportunities for women overall has been slow to open for African American women, opportunities on all athletic administration levels have been more limited (Wicker, 2008). Many African American women are finding other entry points to career in athletics administration (Wicker, 2008). Research shows that while the number of African American women serving as head coaches has been dismal, positions such as Compliance Coordinator/Director, Academic Advisor, Life Skills Coordinator, and Ticket Manager are being filled at a higher rate than head coaches and ADs (NCAA, 2007–2008). Thus, the traditional career pathway from coaching and teaching to administration may not be the conventional path for African American women (Wicker, 2008).

Tiell (2004) studied the career paths, roles, and tasks of Senior Women Administrators (SWAs). This position is the highest ranking position held by a woman in every member institution in the NCAA. This study found that over 58% of the SWAs had prior head coaching experience before advancing to administration. Teel (2005) interviewed 48 current female ADs at NCAA Division I and II institutions and found similar results and additionally concluded that the majority of the respondents were once high school and collegiate student athletes. These studies suggested that the experiences gained, specifically, from coaching and teaching on the college level were integral to the ascension to leadership positions within college athletics (Wicker, 2008). Thus, throughout the years the transition of women from coaching to administration was advanced by the leadership of the NCAA (Wicker, 2008). Legislation, such as Title IX, mandated a more equitable society in the classroom and in the field of play for women in sports, and individual women continued to advance (Wicker, 2008).

Rolle et al. (2000) explored the experiences of eight individual African American administrators at predominantly White higher education institutions in the southeastern United States. The African American administrators wishing to pursue an administrative career in predominantly White institutions (PWIs) find little in the professional literature to prepare them for their journey. Four themes were determined from their study: (i) racial, (ii) self-assurance, (iii) good communication skills, and (iv) and political abilities in higher education. The participants were asked what advice they would give to future African American administrators at PWIs, they said that academic background in general humanities, strong work ethic, and effective communication skills are important skills needed to work in administration.

As women's sports grew in stature, men became more interested in coaching and administering them (Radlinski, 2003). Coaches and official for women were beginning to be paid for their work. More men historically have been involved in athletics and from this group is a large contingent of men who are skilled in coaching. With women's sports coaches paid on a comparable basis to coaching men's sports, the men have found another opportunity of employment in such coaching. Today, few women's athletic programs remain separate from the men's program with females directing less than 16% of NCAA women's sports programs and 21.3% female ADs (Sawyer, 1992).

Gribble's (2011, June 09) article stated:

> Because of title ix legislation most schools merged their athletic programs during the 1970s and '80 s. but in recent years, university of arkansas merged its two departments under similar circumstances and university of texas just combined their programs in 2012 with tennessee still being separate. ("ut merging men's, women's athletics," para 8)

This section summarizes what research exists on the career paths of athletic administrators by the same diversity groups profiled above in the section on the problem of practice: White women, minority men, and minority women.

Career Paths of Women in Collegiate Athletic Administration

Research on the career paths of women in athletic administration has focused on the barriers that they face may be related to gender-specific barriers (Radlinski, 2003). Harriman (1996) describes barriers that appear in two stages: (i) women's entry into the organization and limitation of access to appropriate jobs and (ii) upward mobility limited by the glass ceiling.

Studies have described barriers to advancement ranging from lack of mentoring and networking, lack of career development, gender roles, and background experiences in human capital that women in athletic administration have found to be important. The updated report on the status of women in athletics based on Acosta and Carpenter's (2014) longitudinal studies is 22.3% of ADs are female. The 22.3% figure is the highest representation of females as ADs. What the research does not probe is how women fare in the ladder of positions leading to the AD position, or whether women follow the same ladder as do men.

Career Path of Minority Males in Collegiate Athletic Administration

The literature on the career paths of minority males in athletic administration is exceptionally sparse. In short, Myles (2005) provides empirical research on this topic, and his study focused on the absence of color in athletic administration at Division I institutions. Based on her research, Myles advocated that if Blacks need to scale the career ladder and break through the glass ceiling to dismantle the monopoly on positions in athletic administration. The NCAA has recognized that the opportunities for Blacks to get into athletic administration and advance to top positions are limited. In 2001, the NCAA created the Leadership Institute for Ethnic Minority Males to specifically address the issue of the low number of minorities in senior-level positions in athletic administration. The program was designated to train and educate Black males currently in athletic administration, but seeking to move into these positions. The program is an intensive 12-month leadership training and skills development experience. The goals of the institute are to enhance job-related competencies in the areas of leadership and administration, human resource management, financing, fundraising and boosters, and public and media relations. Participants are nominated by their respective institutions and selected by the Minority Opportunities and Interests Committee (MOIC) of the NCAA. The MOIC selects no more than 25 minority males to participate in the program each year. With the exception of the contributions made by Myles, little is known about the career paths of minority males in athletic administration.

Career Path of Minority Women in Collegiate Athletic Administration

Recent literature on women in athletic administration has touched on issues relating to the career paths of minority women in athletic administrations (e.g., Acosta and Carpenter, 2004; Grappendorf and Lough, 2006; Teel, 2004; Tiell, 2005). However, there is little research that has focused on factors that have contributed to the disproportionate numbers of African American women in athletic administration in the NCAA. Wicker (2008), focusing on Black women in athletic administration, argued that there is a need to explore the career pathway factors and their impact, as well as those factors which have enabled entry and advancement for African American women in NCAA athletics administration. The same holds true for minority women in general.

Several empirical studies have been conducted based on demographic, educational, and career characteristics of ADs. The various experiences of the ADs stem from the predominantly White colleges and universities and the historically Black colleges and universities (HBCUs), in which, they are compared and analyzed. In a study of ADs of historically Black colleges, data was collected and compared to data from previous studies on ADs of predominantly White colleges and universities in which, Quarterman (1992) identified career experiences leading to the AD position. In all, those ADs studied who at HBCUs had been collegiate athletes (76.3% of the 55 respondents) and coaches of high school or collegiate sport (89.0%) (Quarterman, 1992). Moreover, the ADs at NCAA Division IA, now referred to as the Football Bowl Subdivision institutions, came through the ranks as assistant or associate ADs (Fitzgerald et al., 1994).

Herron (1969) conducted one of the earlier investigations of 460 ADs at PWIs that were randomly selected by a recognized lottery procedure. Findings from the study reflected that the majority of ADs across the three levels (i) held a master's degree; (ii) held a degree in physical education; (iii) had participated in basketball, football, baseball, and track at the interscholastic and intercollegiate levels; and (iv) had secondary teaching and coaching experiences that were highly significant factors in the selection process for becoming a collegiate AD (Quarterman, 1992). Further, the study revealed that secondary teaching and coaching experiences were highly significant factors in the selection process for becoming a collegiate AD (Quarterman, 1992). Secondary-school athletic directorship was the specific area in which most of the ADs had gained administrative experience prior to their collegiate positions (Quarterman, 1992).

Williams and Miller's (1983) study explains further that a random sample of 320 collegiate AD (163 males and 157 females) participants were examined based on information relative to demographics, job duties, background experiences, and course work undertaken. The male ADs averaged 47 years of age, and the female ADs averaged 40 years. Most ADs held a master's degree (88%), 5% held a doctorate, and 7% held a bachelor's degree. Half (51.5%) of the ADs also had teaching responsibilities, and 41.5% coached athletic teams. Ranking of job responsibilities was significantly affected by the competitive level or division of the program administered, but not by the gender of the AD or whether the AD headed an NCAA or Association for Intercollegiate Athletics for Women (AIAW) program (Williams and Miller, 1983). Regardless of the gender of the AD or of the program division, graduate course work experiences given the highest ratings were those of communication skills, business-related skills, and public relations followed by knowledge and skills unique to athletic administration (Williams and Miller, 1983). There were significant gender differences, but no division differences in both the percentage of ADs who had certain coaching, competitive, and administrative background experiences, and in the ADs' ratings of how beneficial these experiences were (Williams and Miller, 1983). Although the overall results of the study did show some differences in competencies favoring male ADs, the differences were not such as to support the almost total stereotyping of male ADs in head positions of combined programs (Williams and Miller, 1983). The need to redirect efforts toward improving opportunities for women to gain access to top management positions appears warranted (Williams and Miller, 1983).

Hatfield et al. (1987) solicited the responses of 58 ADs of NCAA Division I programs and of 62 general managers (GMs) within professional sport programs regarding demographic information, perception of job responsibilities, and educational recommendations. Prior to becoming an AD, 70% held positions as head coaches, 48.2% as associate ADs, 36.8% as business managers, and 15.5% as physical education chairpersons. The GMs rated the areas of labor relations and personnel evaluations as more important, while the ADs assigned higher ratings to all other categories such as marking, financial management, administration, and public relations (Hatfield et al., 1987). A discriminant function analysis upon the individual job items corroborated these differences (Hatfield et al., 1987). This study attempted to characterize two of the top management positions in athletic administration and to generate information for curricular planning purposes (Hatfield et al., 1987). The findings underscore the need for graduate programs to provide specialty tracks to serve such different positions (Hatfield et al., 1987).

Landry (1983) investigated through a survey, the demographic, educational, and background experiences of 50 ADs, of NCAA Division IA status from the most competitive athletic conferences. This investigation revealed that the majority of the ADs was outstanding participants in football, basketball, baseball, and track in high school and college and won special recognition such as all-Conference or all-America. They attended a college that had an enrollment of over 15,000 students and received their master's degree and later taught health and physical education in high school. Nearly half (47%) had coaching experience as an assistant or head coach. They averaged 41 years of age when first selected to the position as AD, (i) none currently had teaching or coaching responsibilities, and (ii) they averaged $50,000 in annual salary.

Cuneen's (1988) conducted a study of NCAA Divisions I and II ADs and collected demographic data pertaining to age, sex, administrative experience, sports coached, sports played, degrees held, and major fields of study. The purpose of this study was to design a curriculum for graduate-level preparation of NCAA Divisions I and II ADs.

Findings from this investigation showed that (i) 60% of the ADs were between ages 46 and 50, (ii) 72% had come through the ranks as an assistant or an associate AD prior to becoming the AD, (iii) 58% held master degrees, and (iv) physical education or a related field was the most common academic major. Cuneen's (1988) study of NCAA Divisions I and II ADs, 72% had been assistants or associates before assuming the AD position. It was concluded that a graduate curriculum to prepare a collegiate director of athletics should be implemented through the collaborative effort of an interdisciplinary faculty and that the program should culminate with a doctoral degree.

C. CAREER PATHWAYS

Another investigation by Terry (1988) described the career paths, past experiences, and educational background of chief athletic administrators at 175 small, private colleges and universities in the southeastern United States. A survey was conducted in an attempt to assess the characteristics and professional development of the person serving as AD. Only institutions that were private granted a bachelor's or master's degree and had a total enrollment of 3,500 students or less were included in the survey (Terry, 1988). Athletic administrators at 120 (68.5%) institutions returned usable questionnaires (Terry, 1988). In this study, ADs averaged 46.1 years of age and 9.5 years as ADs. Most (62.2%) held a master's degree as their highest degree, 29.3% held a doctorate, and 8.5% held a bachelor's degree as their highest earned degree (Quarterman, 1992). Undergraduate degrees in health and/or physical education were held by 69% of the ADs. Also, half (50.4%) held graduate degrees in health and/or physical education (Quarterman, 1992).

Furthermore, three most recent studies have shown African Americans in chief athletic administrative roles. Truiett-Theodorson (2005) examined the career patterns of African American ADs at predominantly White higher education institutions. There were factors identified which facilitated advancement within athletic administration, in addition to factors which were perceived to have hindered advancement. Myles (2005) explored some of the challenges that stymie Blacks from breaking into athletic administration and rising to decision-making positions in athletic administration. Five factors were identified that limited Blacks from entering the profession of athletic administration and advancing in the profession: (i) stereotypical beliefs that define Blacks possessing inferior leadership skills and thinking capacities as they relate to athletic administrative positions as they do in society as a whole; (ii) discriminatory acts defined as actions taken to favor Whites over Blacks in securing athletic administrative positions and also actions that limit the professional progress of Blacks in athletic administration; (iii) racist attitudes were defined as deeply ingrained attitudes against Blacks because of the color of their skin that systematically and systemically hinders Blacks from consideration for athletic administrative positions; (iv) old boys' network which was defined as mostly White men who are interconnected across the athletic administration profession and are extremely resistant to hiring Blacks for positions in athletic administration they deem exclusively for Whites; and (v) positional segregation defined as the streamlining of Blacks into certain positions in athletic administration that limit advancement to higher positions and offer very little acknowledgment and benefit. The results found that stereotypical beliefs, discriminatory acts, and racist attitudes were no longer primary factors limiting Blacks from entering the athletic administration profession or advancing in the profession. Although these factors were still relevant, they were not deemed applicable to the overall concern of the lack of color in senior-level athletic administration positions in Division I institutions. The old boys' network and positional segregation were two factors that participants perceived had a significant impact in limiting Blacks from entering the athletic administration profession and advancing. Participants also cited the lack of mentoring as a significant factor.

Finally, Wicker's (2008) study examined the lived experiences of 10 African American women who have succeeded as leaders in athletic administration in the NCAA. The findings revealed that these women acquired life lessons through both formal and informal education which ultimately impacted their career paths. All of these women excelled as student athletes and entered the profession as coaches and teachers that served as opportunities/entry points to athletic administration. Three major conclusions were derived from the findings: (i) formative experiences such as family's educational expectations influenced their career development; (ii) their career pathways to athletic administration were enhanced through mentoring and formal professional development training; and (iii) the good old boy network and race and gender discrimination were identified as perceived barriers to career development.

Although none of the aforementioned investigations targeted ADs of HBCUs, Terry (1988) found that ADs of predominantly Black institutions had served longer in athletic administration than their White counterparts who had studied at small, private colleges and universities in the southeastern United States (Quarterman, 1992). HBCUs are "institutions established before 1964 and have a principal mission that was and is the education of Black Americans" ("Strengthening Historically," 1987, p. 30536).

Although the field of athletics has experienced rapid growth and change, the underscoring need for updating knowledge about effective administration of these programs at HBCUs and/or women because limited information available on the ADs are well over due (Williams and Miller, 1983; Quarterman, 1992). The contribution of this change has stemmed from the impact of the working forces of the AIAW and such governmental legislation as Title IX (Acosta and Carpenter, 1985). A number of studies have chronicled the changes. These significant changes are in both participation and employment opportunities in sports for girls and women (Acosta and Carpenter, 1985). With the passage of Title IX, a massive growth in participation took place in women's athletics (Acosta and Carpenter, 2014). Leadership positions such as coach, AD, and official, previously mostly held by females, became more frequently occupied by males, yet no concomitant increase in the representation of female leaders in men's athletics took place (Acosta and Carpenter, 2004).

C. CAREER PATHWAYS

The progressions to an AD position is a career path that is possible for women and especially SWAs, however opportunities are limited, and the nature of the position is not always desirable in individuals (Tiell, 2004). But, it remains the same that only 22.3% of women administer women's athletic programs with a surmountable increase in women's athletic participation (Acosta and Carpenter, 2014). While Title IX has opened the doors for women as participants in sports, it has actually caused a decrease in the numbers of women in sports administration, thus causing a baffling inverse impact.

All preceding studies share commonalities in their findings. They all relate to identifying the various career paths of becoming an AD. The characteristics and the data collected from these studies may be helpful beyond simple description and comparison (Quarterman, 1992). It can be utilized to develop a profile for data-based research because the research in sport administration does not exist abundantly and specifically, regarding the research of ADs of HBCUs (Quarterman, 1992). These data are applicable for guiding students who aspire to become ADs in making constructive preliminary curricular and career decisions as they progress through a sport administration program (Quarterman, 1992).

Issues relating to minorities and career pathways in the following studies of Truiett-Theodorson (2005), Myles (2005), and Wicker (2008) continue to explore the challenges and barriers in athletic administration even though there has been an increase since 1995. The underrepresentation of minorities still exists in NCAA institutions. While statistics show women are well represented as student athletes in the NCAA, disproportionate absences of women within the hierarchies of NCAA athletic management and decision making continue to persist, specifically, African American women who have expanded at much slower rates (Wicker, 2008). Even though the numbers of Blacks are smaller than their White colleagues in decision-making positions, Blacks are not progressing from one position to another as fast as Whites (Myles, 2005). If a Black athletic administrator spends many years as an associate AD that person may not be recruited into an AD's position at another school (Burdman, 2002). This is the problem that many of the Blacks in decision-making positions face in trying to advance, and it is even harder for Blacks in lower athletic administrative positions to progress (Myles, 2005).

PART TWO: RECRUITING AND RETAINING WOMEN AND MINORITY JUNIOR FACULTY IN THE ACADEMY

Career opportunities and administrative roles undergoing redefinition in athletics highlight the unique conditions and need for research that will advance and strengthen the preparation of leaders in intercollegiate athletic programs (Williams and Miller, 1983). Perceptions of AD may be identifiable through a portfolio of knowledge essential for the preparation of athletic administrators, although no specific academic discipline or department is mandated as being most appropriate for the housing of a graduate program of studies (Williams and Miller, 1983). Emphasized in an interdisciplinary program for all ADs, regardless of the type of program or whom it serves, should be the development of skills in communication, public relations, business, and skills unique to athletic administration (Williams and Miller, 1983). Although it is not necessary to provide different preparation patterns for NCAA and AIAW program administrators, job responsibility data give credence to the desirability of providing specialized emphasis areas which reflect the expertise required in different divisional (I, II, III) program structures (Williams and Miller, 1983).

PART THREE: UNIVERSITY STRATEGIES FOR ADVANCING WOMEN AND MINORITIES

Women, who have aspirations for an athletic administrative career, continue to seek support and role models. Various organizations have professional development programs in place for women. Some are the NCAA's Fellows Program and Internships, Black Women in Sports Foundation's Next First Step Program, and the NACWAA/HERS Institute. These organizations have conferences that women administrators attend for additional information. But, many women begin their careers as physical education teachers and coaches before becoming administrators. Others may come from various avenues in the work place.

Strategies that are developed to help heterosexual, White middleclass women obtain athletic leadership positions may not be adequate for women who experience heterosexism, racism, and/or classism (Delano, 1990). Researchers who cite barriers at the individual level often suggest strategies for change that are geared to help aspiring women become better prepared or more qualified by increase in women participation at workshops, attendance at coaching clinics, and membership in professional organizations (Hart et al., 1986; Acosta and Carpenter, 1988a). Inasmuch as those who hire athletic leaders should have the same expectation for men and women and should be encouraged or pushed to embrace change by hiring qualified women for coaching, officiating, and sport administrative positions (Delano, 1990).

The strategies designed to untangle these roots will be the most useful for social change since women are diverse beings who face diverse barriers depending on their particular life situations, multiple strategies must be developed (Delano, 1990). Only when strategies reflect situational differences and the complexities and depth of the problem will there be an increase in the number of women with diverse backgrounds in athletic leadership positions (Delano, 1990).

PART FOUR: RISING IN THE ACADEMY SUCCESSFULLY

The highest representation of female ADs since the mid-1970s is 22.3% as of 2014. This represents increases from 18.6% in 2006 and 21.3% in 2008. In 1972 when Title IX was enacted, females served as ADs in over 90% of programs for women. Division III schools have the highest percentage of female ADs at 30.3%. Some schools have no female, at any level, in the athletics administrative structures. The percentage of schools totally lacking a female voice has increased from 11.6% in 2008 to 13.2% in 2010 but dropped again to 11.3% in 2014. The most common administrative structure is composed of three administrators: a male AD and one female assistant/associate and one male assistant/associate (Acosta and Carpenter, 2014).

PART FIVE: SUMMARY AND RECOMMENDATIONS

There is little diversity among athletic administrators, and even less diversity at the AD level. Some areas of diversity have actually decreased over the last decade as separate male and female athletic operations were merged. While the overall size of athletic administration has grown, diversity has not advanced faster than this growth. Thus, the problem of practice address is the lack of diversity in athletic administration, especially at the AD level, within colleges and universities.

To fully appreciate the AD position, it is important to understand the diversity that exists (or does not exist) currently in these positions. The information provided looks at diversity in terms of women, minorities, and minorities in sports administration. Tables 1 through 11 provide data portraits of the athletic directors (NCAA, 2007–2008).

References

Abney, R. (1988). The Effects of Role Models and Mentors on Career Patterns of Black Women Coaches and Athletic Administrators in Historically Black and Historically White Institutions of Higher Education. Ph.D. dissertation, University of Iowa.

Abney, R., 2000. The glass-ceiling effect and African American women coaches and administrators. In: Brooks, D., Althouse, R. (Eds.), Racism in College Athletics, second ed. Fitness Information Technology, Morgantown, WV, pp. 119–130.

Abney, R., Richey, D., 1992. Opportunities for minority women in sport—the impact of Title IX. J. Phys. Educ. Recreat. Dance. 63, 56–59.

Acosta, V., Carpenter, L., 1985. Women in sport. In: Shu, D., Segrave, J., Becker., B.J. (Eds.), Sport and Higher Education, first ed. Human Kinetics Publishers, Champaign, IL, pp. 313–325.

Acosta, V., Carpenter, L., 1988. Perceived Causes of the Declining Representation of Women Leaders in Intercollegiate Sports—1988 Update. first ed. Brooklyn College, Brooklyn, NY, Unpublished manuscript.

Acosta, V., Carpenter, L., 2004. Women in Intercollegiate Sport. A Longitudinal Study—Twenty-Nine Year Update. first ed. Brooklyn College, Brooklyn, NY, Unpublished manuscript.

Acosta, V., Carpenter, L., 2008. Women in Intercollegiate Sport. A Longitudinal Study—Twenty-Nine Year Update. first ed. Brooklyn College, Brooklyn, NY, Unpublished manuscript.

Acosta, V., Carpenter, L., 2014. Women in Intercollegiate Sport. A longitudinal Study-Thirty-Seven Year Update. Brooklyn College, New York, Unpublished manuscript.

Burdman, P., 2002. Old problem, new solution? Can programs such as the NCAA's leadership institute for ethnic minority males boost the numbers of Black head coaches, athletic directors? Black Issues High. Educ. 19, 24–28.

Bond, 1983. A Comparative Study of Career Patterns of Black and White Higher Education Administrators in Similar Positions and Institutions. The Pennsylvania State University, University Park (Unpublished doctoral dissertation).

Coakley, J., 2001. in Society: Issues and Controversies. seventh ed. McGraw-Hill, New York, NY.

Coakley, J., 2004. Sport in Society: Issues and Controversies. nineth ed McGraw Hill, NY.

Coakley, J., 2009. Sport in Society: Issues and Controversies. tenth ed. McGraw Hill, NY.

Cohen, G., 2001. Women in Sport: Issues and Controversies. second ed. NAGWS an association of AAHPERD, Reston, VA.

Cuneen, J., 1988. A preparation model for NCAA Division I and II athletic administrators. In: North American Society for Sport Management, Calgary, AB, Canada.

Danylchuk, K., Pastore, D., Inglis, S., 1996. Critical factors in the attainment of intercollegiate coaching and management positions. Phys. Educator. 3, 137–145.

Delano, L., 1988. Understanding Barriers That Women Face in Pursuing High School Athletic Administrative Positions: A Feminist Perspective, PhD Thesis. The University of Iowa, Iowa City, IA.

Delano, L., 1990. A time to plant—strategies to increase the number of women in athletic leadership positions. J. Phys. Educ. Recreat. Dance. 61, 53–55.

Deller, J., 1993. The careers of female directors or intercollegiate athletics and the identification of factors considered important in attaining the position of director of intercollegiate athletics. University of Toledo (Unpublished doctoral dissertation).

Dunning, E., 1986. Sport as a Male Preserve: Notes on the Social Sources of Masculine Identity and Its Transformations. Theory, Cul. Soc. 3 (1), 79–90.

Dunning, E., 1999. Sport Matter: Sociological Studies of Sport, Violence, and Civilization. first ed. Rutledge, London.

Feminist Majority Foundation and New Media Publishing Inc, 1995. Empowering Women in Sports: The Empowering Women Series No. 4. Available from: <http://www.feminist.org/research/sports4.html> (accessed May 31, 2009).

Fitzgerald, M., Sagaria, M., Nelson, B., 1994. Career patterns of athletic directors: challenging the conventional wisdom. J. Sport Manage. 8, 14–16.

C. CAREER PATHWAYS

Grappendorf, H., Lough, N., 2006. An endangered species: characteristics and perspectives from female NCAA Division I athletic directors of both separate and merged athletic departments. Sport Manage. Rel. Top. J. 2 (2), 6–20, Retrieved from <http://thesmart-journal.com/endangered%20species.pdf>.

Gribble, A. (2011, June 09). UT merging men's, women's athletics: Cronan to be interim vice chancellor/sports director until June '12. Retrieved from <http://www.govolsxtra .com/news/2011/jun/09/ut-merging-mens-womens-athletics>.

Hart, B., Hasbrook, C., Mathes, S., 1986. An examination of the reduction in the number of female interscholastic coaches. Res. Q. Exerc. Sport. 57, 68–77.

Hatfield, B., Wrenn, J., Bretting, M., 1987. Comparison of job responsibilities of intercollegiate athletic directors and professional sport general managers. J. Sport Manage. 1, 129–145.

Hay, R. (1986, June). A proposed sports management curriculum and related strategies. Paper presented at the meeting of the North American Society for Sport Management, Kent, OH.

Hay, R., 1986. Proposed sports management curriculum and related strategies. In: North American Society for Sport Management, Kent, OH.

Herron, L., 1969. A Survey to Compare the Educational Preparation and Related Experiences and Selected Duties of Collegiate Athletic Directors, PhD Thesis. University of Utah, Salt Lake City, UT.

Heilman, M., 2001. Description and prescription: How gender stereotypes prevent women's ascent up the organization ladder. J. Soc. Issues. 57 (4), 657–674.

Holland, J., 1959. A theory of vocational choice. J Couns. Psychol. 6, 35–45.

Kjeldsen, E., 1990. Sport management careers: a descriptive analysis. J. Sport Manage. 1, 121–132.

Lapchick, R., 2009. The 2008 Racial and Gender Report Card: College Sport. The Institute for Diversity and Ethics in Sport, Orlando, FL, Retrieved from <2008_college_ sport_rgrc.pdf>.

Landry, D., 1983. What makes a top college athletic director? Athletic Admin. 18, 20.

Leong, F., 1995. Career Development and Vocational Behavior of Racial and Ethnic Minorities. Erlbaum, Mahwah, NJ.

Lough, N., 2001. Mentoring connections between coaches and female athletes. J. Phys. Educ. Recreation Dance. 72 (5), 30–33.

Luzzo, D., 1996. Exploring the relationship between the perception of occupational barriers and career development. J. Career Dev. 22 (4), 239–249.

Morrison, A., White, R., Van Velsor, E., 1987. Breaking the Glass Ceiling. Addison-Wesley, Reading, MA.

Myles, R., 2005. The Absence of Color in Athletic Administration at Division I Institutions, PhD Thesis. University of Pittsburgh.

National Collegiate Athletic Association, 1989. NCAA study on women in intercollegiate athletics: perceived barriers of women in intercollegiate athletics careers. NCAA, 9.

National Collegiate Athletic Association, 2008. Ethnicity and gender demographics of NCAA member institutions' athletic personal. NCAA.

National Collegiate Athletic Association, 2009. NCAA ethnicity and gender demographics of NCAA member institutions' athletic personal. Careers. NCAA.

Parham, T., Austin, N., 1984. Career development and African Americans: A contextual reappraisal using the nigrescence construct. J. Vocational Behav. 44, 139–154.

Parsons, F., 1909. Choosing a Vocation. Houghton-Mifflin, Boston.

Radlinski, A., 2003. Women in Athletic Administration in Community Colleges: Identification of Career Path, Strategies and Competencies Found in Preparation for Leadership Roles in Athletics. Central Michigan University (Unpublished doctoral dissertation).

Rolle, K.A., Davies, T.G., Banning, J.H., 2000. African-American administrators' experiences in predominantly White colleges and universities. Community Coll. J. Res. Pract. 24 (2), 79–94.

Quarterman, J., 1992. Characteristics of athletic directors of historically black colleges and universities. J. Sport Manage. 6, 52–63.

Sawyer, T., 1992. Title IX: Some positive changes have occurred. J. Phys. Educ. Recreation Dance. 63 (3), 14.

Spilerman, S., 1977. Careers, labor market structure, and socio-economic achievement. Am. J. Sociol. 83 (3), 551–593.

Smith, Y., 1991. Issues and strategies for working with multicultural athletes. J. Phys. Educ. Recreation Dance. 62, 39–44.

Strengthening Historically Black Colleges and Universities Program and Strengthening Historically Graduate Institutions Program; Final Regulations, 1987. Federal Register. 52, pp. 30535–30543.

Sue, D., Sue, D., 1990. Counseling the Culturally Different. Wiley, New York.

Super, D., 1953. A theory of vocational development. Am. Psychol. 8 (4), 185–190.

Teel, K., 2005. A Study of the Female Athletic Directors at NCAA Division I and Division II Institutions. Baylor University, Unpublished doctoral dissertation.

Terry, S., 1988. The private college athletic administrator. Athletic Admin. Off. Publ. Natl. Assoc. Collegiate Directors Athletics. 23, 551–593.

Theberge, N., 1983. Towards a feminist alternative to sport as a male preserve. In: NASSS, St. Louis, MO.

Tiell, B., 2004. Career Paths, Roles, and Tasks of Senior Women Administrators in Intercollegiate Athletics. first ed. Pro Quest Information and Learning Company, Ann Arbor, MI.

Tiell, B., 2005. Gender Career Paths in Intercollegiate Athletics. Tiffin University, Tiffin, OH, Retrieved from < bruno.tiffin.edu/btiell/WLS/SWA%20Research/Gendered%20Career%20paths.ht m >.

Truiett-Theodorson, R., 2005. Career Patterns of African-American Athletic Directors at Predominantly White Higher Education Institutions: A Case Study, PhD Thesis. Morgan State University, Baltimore, MD.

Wicker, I., 2008. African American Women Athletic Administrators: Pathway to Leadership Positions in the NCAA: A Qualitative Analysis, PhD Thesis. North Carolina State University, Raleigh, NC.

Williams, J., Miller, D., 1983. Intercollegiate athletic administration: preparation patterns. Res. Q. Exerc. Sport. 54, 398–406.

TRANSITIONING

From Student to Full Professor

Pauline Mosley

Seidenberg School of CSIS, Pace University, Pleasantville, NY

I have always had a passion to teach from a young age. I can recall at 6, propping my dolls and teddy bears on my bed and magically my bedroom transformed into a classroom. The wall became the blackboard and I began to instruct these lifeless creatures on how to add numbers. Many times my grandmother, who was watching me at the time while my parents went to work, would get frighten at all of the noise in the "classroom" because one of the teddy bears was acting up and I had to reprimand them, or Suzy, my favorite doll, just couldn't understand how to subtract so I had to repeat the process again loudly. These classroom sessions would go on for hours with me doing all of the questioning and answering. As I grew older this dream of teaching and the ability to empower students with knowledge became my calling.

When Dr. Susan Merritt, founder and former dean of the Seidenberg School of Information Systems at Pace University in Pleasantville, New York, suggested that I write a book on our way to the airport after presenting at a NYU conference on *Alternate Paths to University Positions* in South Carolina (Merritt et al., 2007); I thought that she was joking! Dr. Merritt, one of my primary mentors, who has played a vital role in my career strongly encouraged me to share my story and perhaps seek others with a similar story to tell. At the time, I couldn't imagine how I would add another thing to my already arm-length list of obligations. In fact, after I told Dr. Merritt that I would seriously consider writing a book, I called my husband to tell him of my next feat. But after a few months of not even knowing where to begin, I called the project off. However, after presenting at several conferences I began to notice that there were very few women attending these computer science conferences and more disturbing even fewer minority women or minorities for that fact in attendance. I asked myself the question, where are they?

Navigating Academia: A Guide for Women and Minority STEM Faculty.
DOI: http://dx.doi.org/10.1016/B978-0-12-801984-9.00008-0.

This continued to trouble me greatly as I attended various conferences in multiple states and even international conferences in Kerkrade, the Netherlands (Doswell and Mosley, 2006a,b), the participation of these groups (women and minorities) was less than 15%. I did further research and discovered that despite persistent recruitment and retainment efforts, women continue to be underrepresented at senior levels. The total number (men and women combined) of doctorates awarded nationally in 1973 was 33,755 (National Science Foundation Doctoral Awards Study, 2005). This study further revealed that of that number, Black women accounted for 0.5%, Hispanic women 0.13%, and Asian women 0.06%. Twenty-five years later, based on data from 1998, when 42,683 doctorates were awarded, the percentages were as follows:

Black women—2.5%
Hispanic women—1.9%
Asian women—5.9%

Unfortunately, gender disparities continue to be a pressing issue within academia. I suddenly realized that writing this book really did matter to me. While I am not a renowned computer scientist, prolific writer of research, nor have I discovered any major technological breakthrough, I have met far too many women who marveled at my life and have asked me the question—"How did you become a professor? How did you do it?" My reply has been, "lots of hard-work and never giving up." I explain that my journey from student to full professor is quite unconventional, but when you have a passion and a strong support network it is amazing what you can do. It has taken me 16 years to attain my current academic position. I am truly blessed to be surrounded by individuals who believe in me and support my efforts, and combined with the ability to take periodic "mental-rests," which allow me to reflect on what I am doing and where I should be going has propelled me to this level of success. I have had many challenges along the way, but, where there is a will there is a way. I truly hope that the reader will be encouraged to follow their passions and those seeking careers in academia will be enlightened and motivated to preserve and help make a difference in the academy.

BECOMING QUALIFIED—GETTING THAT PH.D.

According to data from the Doctor of Philosophy (Ph.D.) Completion Project conducted by the Council of Graduate Schools—the main organization representing graduate institution deans in the United States—fewer than 60% of students entering graduate school in the sciences will complete their doctoral degree within a 10-year time frame. Their recent study conducted in 2010 revealed that about one in five people in the life sciences drop out entirely during the program. And by about year 6, according to the project data, only about 42% of doctoral students in the life sciences will have completed their degree, 34% will still be slaving away at it, and 24% will have thrown in the towel. For the math and physical sciences, only about 39% complete their degree by year 6, 27% are still going, and 34% have dropped out. These numbers are staggering!

One of the biggest challenges confronting women and minorities is earning a doctorate. Most universities and colleges will allow faculty to teach without a degree; however, ascent within the academy cannot begin until one has obtained a Ph.D. Consequently, acquiring a doctorate is a monumental achievement for most women because of the time and cost required. One main reason women or minorities failed to complete the doctorate and remain ABD (All But Dissertation) is the lack of a mentor.

The dissertation process is a maze and can be very confusing. One must understand curriculum, knowledge producing processes and procedures, as well as institution's culture, values, and power relations within and beyond the academy. These are critical success factors. Many potential Ph.D. candidates (abandon the path to advanced degrees) without a mentor to provide direction through the dissertation maze. Others give up because they do not know how to cope with the mechanisms of inclusion and exclusion driven by power relations in the academy. Some students are also inhibited by the challenges faced while writing a dissertation without the support of a mentor and eventually withdraw from the process.

Acquiring two Bachelor of Science degrees, one in Computer Science and the other in Mathematics, from Mercy College was not complicated. After obtaining a Masters degree in Information Systems at Pace University, I began working as an adjunct at various local colleges (Westchester Community College, Mercy College, Iona, The College of New Rochelle, City College, and Hostos Community College) in their Computer Science and Information Systems departments and eventually landed in a position as lecturer in the Business department at the City University of New York, Hostos Community College. This lecturer position was a tenure-track position and did not require a doctorate in order for me to advance, since Hostos Community College was a 2-year institution. I knew that I wished to remain in academia and would not be able to advance much further without obtaining a Ph.D., so my next step was the most difficult undertaking—beginning doctoral studies while teaching full time.

D. TRANSITIONING

CHOOSING A DOCTORAL PROGRAM

Selecting a best fit doctoral program was not a clear-cut process. I just recently gotten married, purchased a co-op, and just landed a full-time lecturer position at Hostos Community College and now I was thinking about pursuing a Ph.D. How would my husband and I finance this? When would I get the time to study? I met with my department and chair and he was very supportive of me returning back to school. As a matter of fact, he suggested that I apply to the City University of New York Graduate Program and because Hostos Community College was a part of CUNY, I could attend tuition-free. This was a godsend! My chair constructed my teaching schedule such that it would allow me to take three courses during the semester while meeting my teaching and office hour obligations. The school provided tuition support and flexible time for the days that I needed to leave work early to get to class, and I was able to continue to work full time as I pursued a degree part time. Managing time was critical to surviving and maintaining sanity. Working full time, grading papers, keeping office hours, taking courses, completing homework and studying, as well as my dual roles as wife and daughter (caring for my elderly parents) presented numerous challenges.

RIGORS OF THE DOCTORAL PROGRAM

I completed first year of graduate studies and my GPA was a 3.6. In my second year of graduate studies, I noticed that study groups were formed and if you weren't in a study group you were at a major disadvantage. Study groups are critical to surviving, especially in my case where my time was so critical. Some groups were privy to tests and resources that other students didn't have and this was another challenge for me. How do I become a member of a study group? The dynamics of how they were formed was most interesting. Similar, to most classroom settings students of similar cultural background or interest tend to form a cohesive coupling. This could be because of the commonalities between the students that establish a bond, thereby they can begin to trust each other with their deficiencies and complement each other with their strengths. In any case, I "pushed" my way into a study group and gained acceptance by volunteering to organize the lecture notes. Because my note-taking skills are rather good, I became an asset to the study group and thus I was in! I strongly recommend joining a study group or creating one of your own, for they are the key in completing the doctoral coursework.

During the second year of my studies, I became pregnant with my first son. The stress of studying, working, and keeping all of the prenatal appointments was too much for me. And not knowing how I was going to juggle a newborn baby in the mix led to me almost losing the baby. Three weeks before the end of the fall semester, I was hospitalized and placed on in a trendenlberg position (head down—feet up) and given strict orders to stay that way until the baby was born. I was 26 weeks along in the pregnancy. My colleagues at work covered my classes for me and a very close colleague and friend who also worked at Hostos modified my final exams and took them to the college to be administered. It is a good practice to seek friendships inside and outside of one's department and school. Prof. Eunice Flemister, an instructor of Gerontology worked in the Health and Sciences department—yet she was my most trusted ally, friend, and confident in the entire college. Although, I no longer work at Hostos, we have maintained our friendship and see each other every week even until today! My husband transported all of the student papers back and forth from the college to the hospital. I graded all of their work and submitted grades from the hospital. However, my professors in the doctoral program did not so understand as well and I was given a incomplete in two subjects and told to withdraw from the other and take the course again next semester, which is what I ended up doing. Although, my focus was on obtaining this degree, lying upside down gave me plenty of time to think and reevaluate my value system and priorities. How important was this degree to me? Was it really worth all of this effort? Our son, Marcus Paul Mosley was born on December 30, 1994, at 2 pounds 10 ounces. It is a miracle that he survived and that today he shows no signs of his prematurity. Two months later, when things settled down I contacted all of my professors and submitted all outstanding work and the incompletes which I received due to the birth of my son were replaced with B's and A's.

The Qualifying Exams

In my third year of doctoral studies, I was confronted with the qualifying exams. These are killer exams. The purpose of the Qualifying Exam is to determine whether a student is ready to begin the Ph.D. thesis research. The preparation for this exam gives a student the opportunity to integrate studies completed during graduate school into the general research discipline of a proposed research area. The test usually consists of two 5-h exams administered in 2 days. The exams cover several areas of study and you can cover two areas on which you would like to be examined. The format and testing procedure may vary based on the Ph.D. program and school. This is where study groups pay off in a major way. Luckily for me, I was off during the summer months from work and was able to devote all of my time on preparing for these exams. The amount of time needed to prepare is crazy. One needs to be extremely focused, disciplined, and mindful of one's time all of the time. This period of my life evolved around studying, studying, studying, and sleeping and eating for about 5 h and then resuming to study, study, more studying. My study group proved to be quite beneficial as we would structure our study sessions and help each other understand complex concepts and ways to remember certain principles, diagrams, and rules of computer science. My husband at first couldn't understand why I was studying so much. He was extremely supportive of my need to prepare and very understanding during this time when meals weren't cooked or laundry wasn't done. However, in order for our household to run smoothly, we had to outsource some chores. Meals were outsource to both my mother and mother-in-law who were extremely instrumental in helping us through this challenging time. The cleaning was outsource to "Molly-The Maid" a cleaning service which we invested in every 3 weeks. Blocks of time were scheduled throughout the week so that I could study and work on papers, assignments, and prepare for upcoming tests. Again, my mother and mother-in-law helped with the babysitting and care of our younger son, Marcus. My parents and in-laws provided us with a very strong support network in which we could leave Marcus with, while I went to study groups or helped in preparing meals and other errands. Without this backbone of support, I could not have made it. We survived and after successful completion of these exams, our lives resumed back to normal for a brief time.

So Close But Yet So Far—ABD

The next phase was the dissertation. This was the biggest hurdle and roadblock in my path. Most Ph.D. programs require students to be a full-time student. The CUNY Ph.D. program was no different, at the time, I couldn't understand why this requirement was mandated by these programs. I begged the chair of the program to please let me continue part time, his answer was an emphatic no. Up to now, I had been taking courses and was matriculated as a part-time student. Becoming a full-time Ph.D. student meant not working. I could not work and attend classes at the same time. This meant that not only would I not have a salary to pay my bills, but I would also incur an additional debt of the courses needed to complete the Ph.D. since my place of employment would no longer be paying for my graduate studies. I tried speaking to dissertation advisors within program to see if they would be my advisor and perhaps speak on my behalf to the program chair and let me begin working on my research while working full time.

All of the advisors refused to work with me because they felt that my time would be divided between work and research and it would be impossible to complete. The faculty and the chair of the doctoral program informed me that a requirement of the Ph.D. program is to be a full-time student, which would require me to leave my job. Furthermore, the faculty wanted graduate assistants to help them with their research and felt that with all of my obligations, I would not be a good candidate. I remember, going home that day and just crying and crying, because I had come so far in the program and now because of money I couldn't continue. The irony of this is that most of the students in the program were hoping once they completed their studies to land in a teaching position in a college. I had a full-time teaching position and was being told to quit so that I could earn a Ph.D.

This is big issue with these programs, especially for working women and women with children. How do you support yourself or your children while attending school full time and you are not allowed to work? I noticed that most of the students enrolled in this Ph.D. program were international students who were on visas or being funded by their countries to get an "American Ph.D." with the intent of returning back to their countries, thus funding was not so much an issue for them. Student stipends and student housing may work for the single student without a house or car notes, but, for the student who is married or has bills to pay this is rather daunting. My husband could not afford to pay tuition, mortgage, and daily expenses. So, in 1997, after completing 60 credits of coursework and taking the qualifying exams with only the dissertation to complete, I joined the ranks of those who are ABD.

Never Give Up—Doctoral of Professional Studies

I knew that I needed a doctoral degree for academic advancement. My options were: to quit my job and attend a Ph.D. program full time and apply for an assistantship, to continue working full time, or find a Ph.D. program that could be pursued part time in the evenings. However, the professional opportunities I was experiencing and my financial commitments precluded me from pursuing a doctoral program full time, thus my options were limited.

In 1999, the School of Computer Science and Information Systems of Pace University initiated a doctoral program in Professional Studies. The Doctor of Professional Studies in Computing (DPS) is an innovative post-master's doctoral program that is structured to meet the needs of the practicing IT professional (Pace University 2001). Unlike traditional doctoral programs that are often narrowly focused, this program emphasizes integrated study between the computing disciplines as well as applied research in one or more of them. It is an intensive, part-time doctoral program designed for completion in 3–4 years. For me, it truly was a blessing!

This innovative degree program addresses the inflexibility of traditional doctoral programs for working professionals. The DPS, while advanced in content and rigorous in its demands, can be distinguished from the Ph.D. in that its focus is the advancement of the practice of computing through applied research and development. The Doctor of Professional Studies is a professional doctorate that integrates academic and professional cultures. The program uses a team approach to both teaching and learning and combines monthly face-to-face weekend meetings with asynchronous distance learning via the Internet. This program promised students a Ph.D. at the end of 3 years, additionally providing the convenience of online courses to students. My husband attended an information session with me and agreed to work much overtime to provide me with the opportunity to seek a doctorate once again.

Coursework Again

Since this was a new program my prior course work could not be transferred to this program so I had to start all over again and take 60 credits of courses. The structure and nature of this program was so unlike my prior experience that I actually enjoyed taking the courses. There were 20 students in the first class and we were more than a study group from the very first class, we were family. Jonathan Law and Patrick Wong, fellow classmates, traveled from California to attend 30 resident weekends over 3 years. The experience of learning with other professionals while earning a doctorate has been most rewarding. We still keep in contact and every year the DPS program has an annual BBQ in which students can return and renew these bonds of friendships forged over intense studying.

The first 2 years were very intense as we did most of our course work online and then met once a month for a weekend of face-to-face classes with our instructors. Friday classes started with a dinner and then class work and a full day of classes on Saturday. The program provided all of our meals which enabled us to bond and relax between classes. Unlike, my prior experience which was highly competitive and at times very, very stressful—the support system provided by the faculty and our peers made it doable. My husband, an electrician, worked many, many nights of overtime and on weekends so that he could afford for me to attend this program. Our parents, both mine and his, were invaluable in helping care for our son Marcus so that I could study and that every now and then we could have a "date-night."

Support of family, friends, or even "paid-support" is critical in an undertaking such as this. I personally know of classmates in both programs who more intelligent than I, but because they lacked support they were unable to complete the program. Two of my closest classmates, their marriages ended in divorce because the intensity of such a program and the dedication and time it demands their spouses just couldn't understand. Another, classmate found it overwhelming trying to juggle work, family issues, personal issues, and studying. Again, the lack of encouragement and the lack of a mentor drove him to relinquish the idea of obtaining a Ph.D. So, I am truly blessed to have such a strong support system to fuel my energy to keep going as well as to have them help me to keep going. Consequently, I completed the coursework smoothly and soon I was ready to begin working on my dissertation.

This time there were many advisors to select from and all very supportive. I determined who would act as my advisor, which is critical to completing in a timely fashion, and decided upon a research topic. My first advisor was brilliant and world renown, and I was humbled to have him as my advisor. However, because of his heavy traveling schedule and many conferences he attended, I found that he was not available as much as I needed him to be. I needed guidance and someone with whom I could meet on a regular basis. Determining who will be your dissertation advisor is critical to you succeeding. I decided to switch advisors and asked Dr. Allen Stix if he would become my mentor. Luckily, for me he agreed. Dr. Stix made the difference in me successfully completion. He was diligent in meeting with me and gave me wonderful suggestions in terms of organizing the research and insuring my accountability to getting certain tasks done.

The Dissertation

Finding an area of research or extending a study which no one else has done is challenging. The next couple of months were spent conducting an exhaustive literary search to understand the "lay of the land." Object Technology was new on the scene and that was the buzz word around town and I was very interested in this type of programming, so I decided this is what I want to write about. Next, finding a focus for research proved to more difficult than I expected.

At first, the appropriate focus seemed to be on what was happening in the first year of programming in response to the addition of classes and objects to the syllabus. Programming with classes and objects complicates considerations of referencing environments and introduces subtleties such as difference between primitive reference and the notion of equality. The derivation hierarchy, method overriding, abstract classes, and interfaces introduce conceptual tool, lots of rules, and more syntax. I wanted to know, how was this all being squeezed in? What was being left out? What impact was this having on the abilities of the students, on their view of the field, and on their enthusiasm? What consequences were rippling over into the second year of the major as students studied data structures and algorithms?

After many conversations with Dr. Stix and his watchful eye I was able to narrow the research down, but, not to narrow and conduct a triangulation study on the cognitive complexities confronting developers using object technology. The next challenge was finding subjects on which to conduct the study. This is where some students end up drowning in data. Dr. Stix wisely advised me that I needed a study that was meaningful but yet required no more than a year to complete. Studies that require enormous amounts of time to conduct are at a higher risk of not getting completed. I conducted these studies on various data sets and was able to finish collecting and tabulating the data within 10 months.

The final phase was writing the dissertation. I love to write, but, what a job! Stating the problem, implications for the current study, research questions to be investigated, external validity, addressing threats to validity, relevance of the research in the context of other work, research methodology,... and more all had to be written. I wrote in the morning, afternoon, night, and way in the middle of the night. When I woke up the next day, I continued writing. It seemed like the writing would never end. After months and months of writing and rewriting and inserting charts and diagrams a dissertation of over 200 pages was completed.

The Dissertation Committee and the Defense

My initial fear of the dissertation defense was quickly abated once I fully understood the functionality of the dissertation committee. The dissertation committee is a concurrent part of any dissertation defense. The role of this committee is to provide you with useful remarks on the dissertation writing, reveal to you any discrepancy in the structure of the dissertation, and offer advice on how to further improve your work. The committee usually consists of three or more members of your choosing. The primary goal of this committee is to help you and to insure that your research is of the highest caliber. Hence, it became crucial for me to select a group of people that were not just knowledgeable in my chosen specialization, but also supportive of my dissertation and my research.

Upon entering the program, I decided it would be beneficial to me to get to know my professors and began thinking about whom I could work well with, who would be supportive of my work, and which group of people would work well together. I selected three people, all of which I knew respected each other as professors and researchers. I selected my chair for his expertise in the field, his prolific writing abilities, and his accessibility. I knew that he would work with me and his goal was for me to succeed. As a focused and disciplined self-starter, I knew that my chair would let me have the space I needed to pursue my research and would not spend his time looking over my shoulder. I trusted him and respected his input highly. Lastly, I knew that our styles would match and he would give me the help and direction I needed when asked.

The other two members of the committee were selected based on their accessibility and their willingness to provide feedback and comments on my chapter drafts. I chose Dean Merritt, because I knew that she would push me to produce my best work and I didn't want to disappoint her. I needed someone who could and would give honest and tough feedback and who would critique closely. The third person, Dr. Mary Courtney, believed in me and assumed a mentoring role and I felt confident that I would trust her as a critic of my work. My dissertation committee was comprised of individuals with varying personal and professional abilities. Consequently each one enriched me from various perspectives as well as provided me with different levels of support.

This was not an easy decision. I spent several months pondering and reviewing faculty profiles for inclusion on my committee. I tried not to listen to the gossip of my classmates, but attempted to focus on what I needed from the committee, what types of personalities I worked best with, and who I thought could tolerate my research and work habits. At first I wanted the top researchers in the field, but, then I began to consider other factors and suddenly I wasn't so sure that I wanted the top researcher on my committee. I decided that what I needed to get through this phase was a group of individuals who would act as my cheerleaders as well as help navigate me through this maze called—the dissertation. I recommend to those students who are at the point of selecting a committee to remember that the committee should cheer critique and be available. Having a cohesive committee will make this difficult part of the program the most enjoyable and rewarding experience of your scholastic pursuit.

Once my dissertation committee met and all members were satisfied with my research. The defense and everything that followed thereafter was mere procedural. The defense date was set and my chair and members questioned me on every possible aspect of the research. I had rehearsed my 70 plus PowerPoint slides so much that I knew it by heart. My adrenaline was pumping and I was ready for any and all questions posed to me from anyone in the audience. I was ready to defend. Up, to this point many of my classmates were writing and preparing for their defenses with the intent of graduating on time. I was extremely nervous that I would not complete the dissertation within the allotted 3-year time frame. During the first year of the program when we were encouraged to produce an "idea-paper" of possible areas of research, I procrastinated thinking... I had plenty of time, so I didn't put much effort into it.

Now, as we are fast approaching graduation I was very, very nervous wondering if I am going to make it. The reality of not finishing and the thought of telling my husband that I needed more time to complete this degree propelled me into a zone of "writing-frenzy." This proved to be quite beneficial, because on April 26, 2002, I defended successfully and was the first one to officially complete the DPS program. I had completed the dissertation within the allotted time frame and with 2 months to spare. It was over! I could now proudly place behind my name DPS.

Becoming a Professor

Like I mentioned earlier, I always had a desire to teach since I was a little girl, I just didn't know in what capacity. My first real full-time job after I graduated from Mercy College was a corporate trainer for Personal Computers Learning Centers (PCLC) located on Park Avenue in the Hemsley building in Manhattan. This was in the 1980s when IBM just came out with their new operating system for the Personal Computer—the Disk Operating System. Hence, an explosion of software applications emerged: MultiMate, DisplayWrite, WordPerfect, Ventura, PageMaker, dBase III, dBase III+, dBase IV, Paradox, Lotus 1−2−3, and host of other software products. PCLC provided training in all of these software applications for all levels of personnel for over 75 major fortune 500 companies. The training would either be 1 day or multiple days. When I first joined PCLC I knew three software programs and did 1-day training for those classes. Over time, I became the training manager and taught instructors how to teach as well as write over 20 software instruction manuals for PCLC.

I enjoyed teaching and I was requested by Pepsico, Brooklyn Union Gas, IBM, Texaco, Dannon, McGraw-Hill, Port Authority, and other major corporations to instruct their CEOs, top management, middle management, and secretarial staff. After a while I grew tired of having new students every single day. The traveling began to take toll on me as my job required me to travel to each of these places to teach. On occasions, the teaching would be done at Park Avenue, but for the most part teaching was done offsite. In time, Brooklyn Union Gas was highly impressed with my teaching style and offered me a job as director of PC training. I accepted and after about 6 months of not teaching just managing the training and sitting in one meeting after another, I realize that this was not the job for me. I left Brooklyn Union Gas and became an adjunct at two colleges.

After a year of being an adjunct, I decided that I was going to become a professor. The local community colleges in which I adjunct at did not require a Master's degree since they were 2-year institutions. However, I was told that if I wanted to be considered for a full-time position I would need the Master's degree. So, I returned back to school to acquire this degree. I continued as an adjunct and learning the ropes of the academy. Once, I obtained my Masters I applied for a full-time position at Hostos Community College (where I was presently an adjunct) and I was hired as a lecturer. My job rank as a lecturer did not require me to conduct research, they were primarily a teaching institution and so my primary duties consisting of teaching and service to the community as well as the college. After 7 years of teaching, I applied for tenure. I was tenured as a lecturer and told that if I wanted to apply for an assistant professoriate rank or higher, I would need a doctorate or evidence of strong research and publications. I decided to pursue a doctorate.

While attending Pace University's doctoral program (DPS), Dean Merritt offered me a lecture position in the Computer Science department. It was a nontenure 1-year appointment position. This was a major cross-road. Do I give up my tenure, which I just recently secured? Or, do I stay put? What do I do? What if I do not impress the CS department enough to extend my contract? What happens after the 1 year? After much discussion with my husband, it was decided that I take a big risk and leave my tenured position and accept a 1-year contract at a 4-year institution.

This was a major change in teaching. I had to revamp my entire teaching style and methodologies of instructing as the student population was very different. The classes were challenging and I was taking classes at the same time and striving to excel in both areas. At the end of my 1-year appointment, I was evaluated and my contract was extended for another year. This continued for 3 years, it was a very tense time for me. Not knowing, if your contract was going to be renewed or not. Thus, I made sure that before each class I was there 15 min ahead of time to prepare and get ready and I stayed after 10 min to clean up the room. I diligently kept office hours and made myself available to the students outside of the office hours. At least once a week, I visited the student labs where students were writing their programs, and offered my help. Dean Merritt advised me to complete the doctorate as quickly as possible, so that I could move from a lecture line to an assistant professor line. In 2002, once I received the doctorate I applied for an assistant professorship and it was granted.

TENURE AND PROMOTION

West-Olatunji (2005) conducted a study focused on the experiences of African American faculty from the framework of cultured-centered theory. This study provided meaning of the Black experience in the academia and further illustrated the issues contributing to the lack of significant numbers of African American faculty in traditional White institutions. Themes were found in the study, which included (i) interaction (bonding) described as little or no effort by colleagues to informally or formally theorize, socialize, or intellectualize; (ii) variables (streams of consciousness) described as Black academics being overwhelmed by the multiplicity of microaggressions enacted by White colleagues in the academic workplace; (iii) no transference of power/authority described as no acknowledgment of Black faculty as real intellectuals by colleagues or students unless there was institutional accountability; (iv) subjective reality of the White experience (reflections) described as the articulated surreality of participants working with their White colleagues despite Eurocentric perspective of investigating the hegemony existing with whiteness and maleness present in academia; (v) mutual benefits of reciprocity and transformation which was defined as a sense of hopefulness that positive outcomes are possible, multiple centers are beneficial, and diunital theorizing creates new possibilities for research and praxis; (vi) disconnections, duality, and divergence entail understanding the effects of oppression in the academic experience; and (vii) resiliency which spoke to participants acts of resilience, self-preservation, creativity, and resourcefulness despite their experiences in the academy.

West-Olatunji (2005) summarizes the study as such:

> Oftentimes, problems of tenure, appointment and promotion of African Americans and other people of color in the academy are articulated as being the result of or the effect of actions of another/others and what those others did or did not do. However difficult for some, it is incumbent upon each person to be responsible for and accountable for his or her own self and actions, especially in such an environment and organizational structure. This is particularly important in a setting such as the academy, where, on one level, each person's duties are independent of others, including meeting with students, lecturing, and contributing to his or her discipline, such as engaging in research and publishing.
> However, on another level, each individual is responsible for "being a good citizen in the community;" that is, one must actively participate on the committees of

the department, the division, the college, and so forth. Because of such an organizational structure, it is imperative that one guards against overextending one's self, so that one is justly balancing all of his or her responsibilities, and producing on all levels. Much of the difficulty lies in being pressured into serving in certain capacities because one is a minority. Minorities in this society and in the academy, are aware of the power and privilege European Americans always have been afforded, so when called upon by these "powerful" colleagues for assistance, many minorities, in an effort to please and to appear amiable overextend themselves as "good citizens in the community," while not having the required time to produce on the other level: so, they do not publish, do not engage in research, and do not perform certain other duties. Much of this has to do with power. Having always been a problem in America, the misperceptions of power, the historical unequal distribution of power [a result of the larger society], and the contemporary "management" of power continue to plague many African Americans in the academy.

Academic quality at universities is conceptualized and defined by tenured faculty who are majority White male and female faculty (Constantine et al., 2008). Patitu and Hinton (2003) discussed the issues some of their study participants discussed with regard to tenure experience and most described more negative experiences. Though some stated there was no problem, the majority voiced concerns. The main concerns that emerged included little or no mentoring throughout the process, being given conflicting information regarding the tenure and review process, higher expectations than their White colleagues, and being subjected to unwritten rules about the process (Patitu and Hinton, 2003).

As a young mother, without a terminal degree, in a tenured-track assistant professor position, I felt overwhelming pressure to perform at a higher level than my counterparts. In addition, I was teaching at the same place where I acquired my Masters and was in the processing of acquiring a doctorate. I remember the feeling, like it was yesterday walking into my office which was next to the faculty whose class I was taking on the weekends. I said, to myself Wow! There is a constant fear of job security knowing that the time clock is ticking and I am expected to obtain a terminal degree in order to maintain my job. I teach four classes per semester, taking 12 credits of doctoral courses, a member of several committees, a volunteer coordinator, while being a wife and mother and juggling the chores that come with being a woman. There is a steady struggle to find equilibrium. The institution of higher education continues to confirm that this male constructed academy was, and continues not to be designed for mothers. There is very limited tolerance for women who decide to have children. My academic duties cannot be affected by maternity leave, a sick child or child care provider, doctor's appointments, or extra curricular activities.

I am extremely fortunate that I did not have to choose between the tenure clock and the biological clock. My place of employment has been extremely supportive and sensitive to women issues and I have been truly blessed to have been mentored by such wonderful women in the academy who readily lend a helping hand to anyone, but in particularly women. For that, I am truly grateful for all the women at Pace University who have helped me along the way! I was spared from these negative experiences and was heavily mentored which made a tremendous difference in whether I would succeed or fail.

Motherhood and Academia

Whether inside the classroom or out, Black women are a minority in higher education. My current institution is highly supportive of women professors. It could be because at the time, we had a female president, several of the deans were women, and my chair was a woman. Thus, their sense of understanding or having gone through various obstacles themselves might have contributed to their sensitivity.

While a majority of Black faculty currently teaches at HBCUs, my current institution still defies the reality of being both Black and female in American Higher Education. I have never seen so many women in leadership positions—Chairs, Deans, Presidents, and Full Professors. I see people who look like me doing great things!

While individual experiences are shaped by institutions, research suggests collectively female faculty with children have greater challenges when it comes to moving up faculty ranks and receiving tenure and promotion. For those with children we are more likely to be in lower ranked faculty positions at research universities or at smaller institutions with a greater teaching and service load. In both scenarios, we are confronted with the demands of academia—University, Department and Committee obligations, service, teaching, student counseling, mentoring, and research.

Motherhood transcends race, age, class, sexuality, and ethnicity yet as women we struggle with the major aspect of our lives in the academia setting, because we don't want to be perceived weak or mindless. When our kids are sick or our parents need care, we struggle with whether to cancel our classes or make other arrangement? How much do we share and will what we share be held against us. The major challenge here is balance between an effective academician and a caring mother or caregiver. For working women, this is a real question because whether married or separated, women do bear a great responsibility for caretaking than men. How has motherhood impacted my career and vice versa?

In reflection, these are a few things that have "shifted" since having two children: My time for deep scholarly reflection and writing occur between 1 and 4 a.m. I negotiate how I will get my son to sleep by 7:30 p.m. and juggle the duties and responsibilities around his schedule.

Do I spend enough time with my son? When I run off to conferences and his babysitters and my husband help him with various projects and homework—do I feel guilty? Yes, am when I can spend time with him—I'm exhausted. Try as you may there really is no ideal balance—you just have to make things work and livable from day to day and not worry about it. You are doing your best and that is all anyone can expect to do. My strategy has become reconceptualizing the meaning of balance. Conflict theory would suggest that there is a madness associated with trying to balance multiple roles because maybe it is impossible. However, I'm willing to live with this madness because I love what I do as a mother, wife, professor, researcher, and educator. So, let the dishes stand in the sink, let his toys lie all over the house, let the laundry pile up, and let the house stay messy for the time being—because my sons and husband know that when summer comes—I am home and things will go back to normal and I have time for them.

Why Aren't There More Female and Minority Professors?

The Answers—according to Dr. Joyce Tang, Department of Sociology in Queens College (Tang, 2007).

> The *short* answer to this question is that fewer women and minorities receive scientific training, and proportionally, greater numbers of women and minorities do not remain in scientific professions. But the "leaky science pipeline" thesis (or the "revolving door" phenomenon) is *not* an adequate explanation for the underrepresentation of women and minorities in science. Despite many legislative changes, women and minorities need to overcome many subtle and no-so-subtle obstacles to enter, remain, and succeed in science. These barriers are cultural, structural, as well as institutional.
> The *extensive* answer can be stated by three postulations, each with its own limitations:
> The *biological* explanation emphasizes women and minorities' lack of abilities to pursue intellectual tasks.
> According to the *individual choice* explanation, women and minorities have a general preference for non-science careers.
> The *structural* explanation argues that institutional obstacles are responsible for the virtual absence of women and minorities in science.

I agree with Dr. Tang, that these barriers are the cause of the shortage and until we begin to understand these barriers and take steps to eliminate them, the shortage will remain. Female and minorities should not let these barriers discourage them nor prevent them from seeking a STEM career—there are solutions and these barriers are not insurmountable! Dr. Tang conducted a sociological study of extraordinary women from different scientific fields, contexts, and countries which revealed that successful scientists can be nurtured, and not determined at birth. Dr. Tang, states "a career in science is a product of individual attributes, structural opportunities, and institutional support. Career advancement is a continuous, dynamic process of choice, design, and adaptation. The right attitude, support, and environment can provide women and minorities with success in science, both academically and occupationally."

Another reason is that African American faculty, unfortunately, have been forced in a revolving door scenario in academia. Often they are brought into majority White institutions and become victim to a lack of effective mentoring, systematically racist institutional climates, and feelings of isolation (Patitu and Hinton, 2003). Consequently, Black faculty experience a devaluing of their scholarship, infringement on their time due to expected campus diversity initiatives and student mentoring, lack of mentoring, as well as constant feelings of isolation and executed racial microaggressions by fellow White faculty resulting in creation of an environment that is less than desirable. These factors thereby cause them to leave or seek a more productive and conducive work environment.

Achieving academic and career successes in science presents a challenge to all of us, especially to women and minorities. The culture and norms of society in general and of science in particular have affected men, women, Whites, and minorities differently, primarily because of socialization and the structure of the scientific profession.

Some women and minorities have become successful because they have overcome numerous obstacles to make significant contributions to the field and the profession. They know how to navigate within the White, male domain. In addition, these women seek a work environment that is committed to building a community to prevent isolation of African American faculty that is practiced at the departmental administration and university administration level (Yoshinaga-Itano, 2006), which are essential key elements to success. Lastly, they work diligently to develop their networks. They take bold and strategic actions to become part of the "old-boys" club. So, ladies and minorities continue to fight, continue to remove these barriers one by one until they no longer exist—because one day, hopefully very soon—the club will be a clubhouse for all.

References

Constantine, M.G., Smith, L., Redington, R.M., Owens, D., 2008. Racial microaggressions against black counseling and counseling psychology faculty: a central challenge in the multicultural counseling movement. J. Couns. Dev. 86 (3), 348—355.
Doswell, J.T., Mosley, P.H., 2006a. Robotics in mixed-reality training simulations: augmenting STEM learning. In: Proceedings of the Sixth IEEE International Conference on Advanced Learning Technologies Conference, Kerkrade, the Netherlands.

Doswell, J.T., Mosley, P.H., 2006b. An innovative approach to teaching robotics. In: Proceedings of the Sixth IEEE International Conference on Advanced Learning Technologies Conference, Kerkrade, the Netherlands.

Merritt, S.M., Dwyer, C., Houle, B.J., Mosley, P.H., Coppola, J., 2007. "Alternate Paths to University Positions: Women's Choice", Faculty Resource Network Journal: Advancing Women and the Underrepresented in the Academy. Available from: <http://www.nyu.edu/frn/publications/advancing.women/Women%20index.html> (accessed March 9, 2014).

Mosley, P.H., Greene, S., Morgan, E., Hargrove, K., Rogers, T., 2010. Why is there a shortage of women and minority faculty? Paper presented at 10th Annual Diversity Challenge Conference—Institute for the Study and Promotion of Race and Culture (ISPRC), Boston College, MA.

National Science Foundation Science and Engineering Awards: "Professional, etc. includes professional, unknown and other", 2005. Percent Female Among Doctorate Recipients.

Patitu, C., Hinton, K., 2003. The experiences of African American women faculty and administrators in higher education: Has anything changed? New Directions for Student Services, 104, 79–93.

Ph.D. Completion Project, 2010. Available from: <https://www.cgsnet.org/phd-completion-and-attrition-policies-and-practices-promote-student-success> (accessed March 9, 2014).

Tang, J., 2007. Advancing women and the underrepresented in the sciences—a sociologist's view. Faculty Resour. Network J. Available from: <http://www.nyu.edu/frn/publications/advancing.women/Plenary%20Tang.html> (accessed March 9, 2014).

West-Olatunji, C., 2005. Incidents in the lives of Harriet Jacobs—a readers theatre: disseminating the outcomes of research on the black experience in the academy. In: King, J. (Ed.), Black Education: A Transformative Research and Action Agenda for the New Century. Lawrence Erlbaum Associates Publishers, Mahwah, NJ, pp. 329–340.

Yoshinaga-Itano, C., 2006. Institutional barriers and myths to recruitment and retention of faculty of color: an administrator's perspective. In: Stanley, C.A. (Ed.), Faculty of Color: Teaching in Predominantly White Colleges and Universities. Anker Publishing Company, Boston, MA.

D. TRANSITIONING

From Faculty to Chair

S. Keith Hargrove

College of Engineering, Tennessee State University, Nashville, TN

There are many challenges confronting higher education in the twenty-first century. They range from cyclical state and federal support, more legislative oversight and accountability, managing limited university resources...to a changing student population, student educational needs, and the need for more effective leadership in academia. One key to addressing these and many other challenges is the continuous development of leadership skills of department chairs, deans, and administrators to find innovative solutions to fulfill university missions, and serve its customers.

Leadership can be defined as the ability to influence others to visualize, understand, evaluate, support, and implement a strategy for transforming an organization to achieve a goal. Leadership and management are not the same. Management is the ability to administer and coordinate policies, procedures, and operations of an organization. It is essential to differentiate and recognize that these are two separate traits or characteristics, though an individual may possess both. After successfully completing the traditional tenure-track route and eagerly accepting more responsibility for administrative activities and functions as an associate professor, I have discovered that leadership is a dynamic activity that requires a number of personal traits and characteristics to be effective and create change in an academic organization.

As an associate professor, my academic growth and experience has been insurmountable in terms of professional development. Serving in the three capacities of teaching, research, and service, I realized that a faculty member actually determines his or her own academic destiny depending on much time and effort that is placed in these three initiatives. Depending on the institution's primary mission, the amount of time you spend on one or the other may differ. However, in my judgment, the key to your success and achievements in any of the three

119

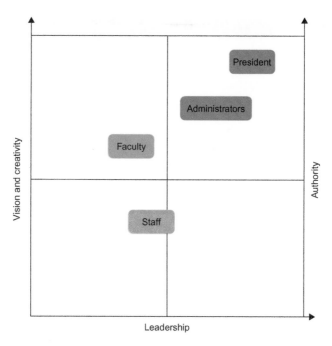

FIGURE 9.1   Academic leadership quadrant.

areas will depend on your ability to "Lead." Within the academic organization, all personnel have specific roles and tasks to support the
President/Chancellor's vision, the Dean's goals, and the Department
Chair's objectives. To accomplish each role, all exhibit certain levels of
leadership, authority, and vision/creativity to make things happen. As
shown in Figure 9.1, imagine a graph with four quadrants with increasing levels or scales of vision, authority, and leadership.

As I reflect on a decade of academic experience, all university personnel should fit in this Academic Leadership Quadrant based on these
three factors. Presidents are required to exhibit the highest form of leadership and vision, with a position of high authority in a university environment (Quadrant I). University administrators carry out the vision of
the president with a lesser level of authority, typically in the same
quadrant as the president. Faculty is usually seen in Quadrant II or IV,
depending on their positional influence and their creativity. Most staff
are in Quadrant III. Though this figure is highly subjective, it may
provide some insight on the levels of leadership impact different university personnel have in academia. My own self-assessment of position and responsibilities has helped me to develop the necessary skills
to gradually move toward Quadrant I. Now after completing my
first year as a recognized and designated administrator (Department

Chair), the purpose of this article is to share my first year experience with others and to offer insights, reflections, hardships, successes, and opportunities.

I began my first year as a Department Chair after serving almost a decade as a regular faculty at another institution. In that capacity, fulfilling the roles in teaching, research, and service has garnered recognition in awards, publications, grants, and service to the community. The decision to become a Chairperson was an easy one: I truly wanted more responsibility for curriculum change and innovation, student's success, and the opportunity to implement my own ideas, whether they were successful or not. I was fortunately granted that opportunity at a different university with some major challenges ahead of me. I would like to share my experience with the reader on my perspectives on the responsibilities of a department chair, and describe my first year experience of developing a departmental strategic plan, assessing the matriculation of students and the curriculum, and implementing an advisory council. I hope the reader will find value in the following pages of text.

TRANSFORMATIONS, ROLES, AND RESPONSIBILITIES OF A CHAIR

Making the transition to department chair should require a careful understanding of the responsibilities, challenges, and expectations of the position. Numerous publications and programs exist to help with this transition (Hecht, 1999; Leaming, 2003; Lucas, 2000), and I would strongly encourage any faculty member considering this career move to investigate these learning and training opportunities. Following is a brief discussion of this transformation.

As in any layered organization, there is a level whose primary responsibility is to develop the mission and goals, and a lower level whose primary responsibility is to implement the mission and goals. In academia, I surmise this position to be a department chair. It is a position that requires a commitment to adhering to administrative policies and procedures, being a change agent, while at the same time serving internal and external customers with all their demands and needs.

Yet, there is a dual challenge in becoming and succeeding as a department chair. One is making the transformation from the traditional faculty role to an administrative role. The second challenge is to do an effective job as chair based on the roles and responsibilities that come with the position. Hecht (1999) discussed the transformation based on some existing models of the department chair. In becoming a chair, recognizing the need to expand one's personal knowledge or expertise to

the general knowledge and familiarity of all faculty, is essential to promote and serve their needs and the department itself. As a former faculty member, your interests and relationships with other faculty was
essentially one-on-one. Now you must assess these personal agendas
and seek their acceptance for a unified purpose in the best interests of
the department. And above all, your personal needs as a former faculty
member are now not as important. Hecht (1998) also concludes that the
attitude of a department chair is just as important to success and effectiveness as serving the needs of the faculty and supporting the university on behalf of the dean and president.

The roles and responsibilities of a department chair should be clearly
defined in the faculty handbook. Their primary responsibility is to manage
and coordinate administrative activities in the department including
curriculum and course offerings, faculty development and evaluation,
departmental resources and budget, set strategic goals, and support the
agenda of higher administrative positions (e.g., dean, vice-presidents,
president). Gmelch and Miskin (1995) and Tucker (1992) discussed the
many tasks and roles of department chairs, and basically outlined four
roles and responsibilities: faculty developer, manager, leader, and scholar.

The department chair must meet the needs of the faculty and provide
resources for their development. This also includes recruiting and hiring
faculty. Because faculty has individual needs, it is a major challenge to
address those needs and concurrently address the collective needs and
objectives of the department. The chair is also faced with personality
issues that may require conflict resolution skills, and seeking out ways
to improve and promote innovative instruction. And lastly, the chair
must provide ways to motivate and reward faculty for their efforts and
recognize them for their accomplishments.

The chair must manage the limited resources of the department while
meeting the needs of the curriculum, students, and faculty. This
involves strict control of the budget. Leadership is defined as the ability
to influence people to work toward a common goal, but a manager is
the person who actually "makes things happen." An effective chair
must have the skills to do both. Hence, the leader sets the vision and
direction of the department. This usually involves developing a strategic
plan with the acceptance and support of the faculty. The challenge is
then to implement the plan to produce the desired results.

The fourth role of the department chair is promoting scholarship,
research, and faculty autonomy. The chair must be seen as someone
who advocates freedom, but also encourages the development of new
knowledge. Seeking opportunities for faculty to obtain research funding
and manipulating the administrative procedures to minimize the
bureaucratic obstacles will help support a chair's vision and goals for
the department. I would encourage all new department chairs to make

an assessment of themselves and the department to fulfill the above responsibilities and challenges.

THE FIRST YEAR EXPERIENCE

A national survey conducted by Washington State University Center for the Study of the Department Chair (1992) concluded that most individuals choose to become department chairs primarily for personal and professional development. This was consistent with my own desire and ambitions. I decided to accept a position of department chair at another institution, rather than awaiting and anticipating for the opportunity at a home institution where my tenure status was obtained. I am sure there are advantages and disadvantages of becoming a new chair at a home institution, or moving to a new environment. I decided to choose the later. The American Council on Education (ACE; 2002) also provided a workshop for new department chairs that I participated in to prepare for my new and upcoming position.

A decision to become a chair requires a serious self-assessment of skills, abilities, experience and knowledge, and the strong desire to seek greater challenges and responsibilities. Though I did have some limited reservations, I was ready for the challenge and the opportunity. After accepting the position and relocating to another university, I became familiar with my prescribed responsibilities as dictated by the faculty handbook, and had one-on-one discussions with each faculty and staff. I decided to follow a three-step adjustment phase to help my transition in learning the responsibilities of the job, gain acceptance and support of the faculty, and begin to develop and implement my own goals and objectives for the department. The three simple phases are described as Assess—Develop—Implement

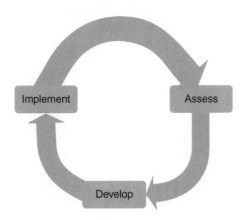

FIGURE 9.2   ADI phases to leadership transformation.

(ADI) as shown in Figure 9.2. The matriculation of the phases is to occur over 3 years. I called this the *Chair's One-Three Plan* in which I would spend the first year (One) listening and learning how the university and engineering school operated. I spent the Fall (Assess) semester primarily learning everything I could about the university including reading all the university manuals, handbooks, etc., to become familiar with my environment. I also completed a departmental assessment of student performance, retention, and curricula based on ABET (Accreditation Board of Engineering and Technology, 2000) engineering criteria. The objective was to determine the "as is" profile of the department. During the next Spring (Develop) semester, I begin to solicit faculty input to begin to develop a 3-year (Three) strategic plan through regular faculty meetings, program reviews, and personal objectives of each faculty and their research goals. Some may extend this period to 5 years. I also recommended a faculty performance criteria and evaluation plan, which was accepted. Each faculty also indicated their research laboratory objectives and needs to help develop the strategic plan. To address student issues and concerns, each semester a department student body meeting was held to obtain feedback from students and inform them on the direction and strategic directions of the department. The department also created an Industry Advisory Council to provide consultation and advice on the curriculum, industry projects, and employment opportunities. The aforementioned activities will be discussed in the following section in the context of developing the strategic plan.

Developing the Strategic Plan

The next phase of the One-Three Plan was developing the Strategic Plan. This process took place during the Spring Semester, though it is dynamic in nature and implementation. The key was trying to determine the goals and objectives of the department, and follow a strategic planning methodology that would lead to a roadmap for growth, dynamic development, supportive change, and progress. Using the ADI process, and a formalized approach to strategic planning, the department created a strategic plan for the next 3–5 years based on current approaches in the literature. These goals and objectives were also juxtaposed to the university and the school of engineering. Elizandro and Matson (2001) describes strategic planning from the perspective of satisfying ABET 2000 engineering criteria that includes formulating goals and objectives, developing the strategic plan, and assessing the effectiveness of the plan. Shelnutt and Buch (1996) used total quality principles for strategic planning and curriculum revision, and Fabiano (2002) examined strategic and operational planning at the department level describing specific steps for implementation. The department also reviewed current reform efforts in Industrial Engineering that are problem-driven versus tool-driven, a dynamic and responsive curriculum, and based on defined core competencies (Kuo and Deuermeyer, 1998) including knowledge, skills, and abilities (KSAs).

Our approach was to extract elements applicable to our interests and environment, and develop a strategic planning model that encompasses the elements of ABET engineering criteria, and the personal objectives of the department. Because the ABET engineering criteria provides such flexibility and autonomy, we believe the model developed in Figure 9.3 provides an approach to achieve our goals in the next 3–5 years. The strategic planning methodology has essentially three major components: Constituents, ABET Requirements, and the Strategic Plan. Our department began this development with regular meetings during the Spring Semester and input was received from the faculty to develop the vision,

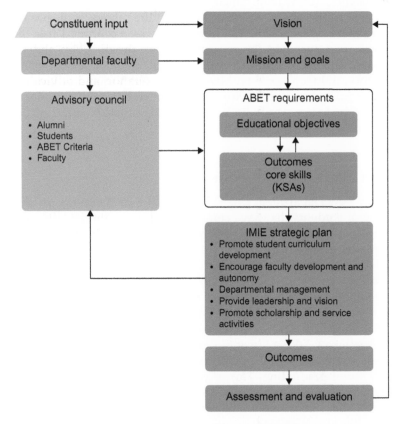

FIGURE 9.3   Strategic planning model for engineering department.

mission, and goals. I believe these elements should come from the faculty...though their input should be based on their knowledge and relationship with industry, research, student and academic experiences, and personal goals for the department. These elements were then exposed and challenged by an Industrial Advisory Council made up of major employers of industrial engineering graduates. The Industrial Advisory Council currently meets at least four times during the academic year, and it also has developed a strategic plan. It will be discussed in Section D. The vision, mission, and goals were adjusted accordingly but satisfying the opinions of the faculty yet promoting an "industry-driven curriculum" (Figure 9.3). The next step was to carefully review the ABET 2000 engineering requirements and make sure they supported the aforementioned vision, mission, and goals, and outlined defined program and educational objectives. Input and feedback was received from the Industrial Advisory Council, departmental Alumni, and even current students by having student body meetings. After meeting these requirements, the department was then able to identify specific initiatives that support the above and provide a roadmap for advancement. The initiatives centered around five thrusts and will be evaluated every semester of progress by the Chair: (i) Student and Curriculum Development; (ii) Faculty Development and Autonomy; (iii) Departmental Management; (iv) Vision and Leadership; and (v) Scholarship and Service. Outcomes were developed for the five initiatives, and they would be assessed and evaluated yearly to support the vision.

Student Matriculation Assessment

As the new Department Chair, I figured the first step was to analyze the type of students that have completed the program, and take a thorough look at the curriculum evolvement over the last 5 years. This information was achieved primarily with discussions with the faculty and completing a retention study for the department. Therefore, the retention study was completed with the objective to determine the graduation rate and the ability to predict the success of engineering students (Hargrove and Domingo, 2003). The departmental study revealed the 6-year graduation rate at 40%, and a minimum high school grade point average of 3.0, SAT scores above a certain range (954), and undergraduate graduation grade point averages of at least 2.5 had a higher probability of graduating. This of course is no surprise to all engineering faculty, but my goal was to determine the criteria for at-risk students from urban environments and for recruiting purposes. This information also provided insight in selecting and scheduling courses for new and existing students in the program if the goal is to increase the graduation rate.

The department has also been faced with a declining enrollment in the last 3 years. Much of this is attributed to lack of recruiting, high school awareness of the industrial engineering discipline, and a summer precollege program to recruit and serve as a feeder for the department. Therefore, an aggressive recruitment effort was initiated and a summer program was developed. The Pre-Engineering Program to Stimulate Interests in Industrial Engineering (PEPSIE) is an 8-week program sponsored by the Lockheed Martin Corporation that focuses on exposing students to the discipline of industrial engineering with hands-on projects in manufacturing, computer-aided design, information systems, energy systems, and real-world case studies. An exit student satisfaction survey was completed by the participants to partially assess the effectiveness of the program. Other recruitment efforts initiated were developing partnerships with selected high schools with pre-engineering and manufacturing programs, having high school student's complete internships in the department, and getting current engineering students involved with the local high schools.

Curriculum Assessment

The next challenge for the department and chair was an assessment of the industrial engineering curriculum. More specifically, did our program prepare students for industry, government, or graduate school, and does the program increase the student's marketability as industrial engineers? A number of reform issues in industrial engineering are being addressed that focus on vertical integration, problem-driven courses, information technology (IT), and the establishment of a core set of competencies (Leonard, 2003; McGinnis, 2002). The department collectively determined the core KSAs for our program, and reviewed the courses based on the methodology proposed by Felder and Brent (2003). In addition, a review and comparative study was completed of existing industrial engineering programs of similar size to assess course offerings, content, and research initiatives. We concluded the program was in need of a basic circuits course and further enhancement of the concentration areas (electives) in the department. Recognizing the employment opportunities in information systems and technology, more development was done in this area with new courses, supplemental faculty from industry, and more IT-focused problems integrated throughout the curriculum. The above activities were completed with the cooperation of students and an Industry Advisory Council. The council will be discussed next.

INDUSTRY ADVISORY COUNCIL

The department revived an existing advisory board that previously appeared to lack direction and focus before my arrival. As chair, I carefully presented the goals and objectives for the department, as well as "how" I believe the advisory council should support the program. With some modifications, we were able to develop a set of goals and objectives for the advisory council with full support. The advisory council now meets at least twice a semester. However, one must maintain that the group exists to provide advice, not as an "executive board." Cutlip (2003) provided one example format for creating and operating a departmental advisory board.

The members of the council consist of industrial and government representatives that have hired our graduates over the last 15 years of the program's existence. There are no term limits, and member participation is voluntary. Though there are more than 20 companies acknowledged as board members, about half have strong members of the department and regularly attend the meetings which include alumni and/or mid-level managers.

The Industry Advisory Council is a partnership of business and government professionals and leaders joined to develop and enhance the curricula and students of the industrial engineering program at Morgan State University. The council has four goals:

1. advise and consult on the continuous development of an industry-driven curriculum;
2. support and engage in industry-sponsored learning activities for students and faculty;
3. provide matriculation support for student success;
4. exist and operate according to established guidelines and an organizational structure.

The council provides advice and consultation on courses and course content in the curriculum. They also help identify KSAs that they would like graduates to possess upon graduation, and have assisted in the student professional development workshops such as interviewing and work ethics. Several members have also participated in an industry—academic lecture series this past year. The council has also made a commitment to identifying summer, permanent, and co-op opportunities for students; student projects; and faculty collaborations. To help with student success, several scholarships have been created by the council, including the sponsorship of the summer program for industrial engineering students. The council is formed in three committees to address the above goals, and have established objectives, strategies, and metrics to accomplish the goals within the academic year. For example, the council completed an alumni survey and a curriculum evaluation to provide feedback to the chair and faculty within the last year based on a former model (Olds and Miller, 1998). The council has played a major role also in developing the strategic plan and helping the department charter a course for the future.

WHAT HAVE I LEARNED ABOUT LEADERSHIP

Before considering, applying, and accepting a position as a department chair, a self-assessment of leadership qualities is important for any faculty. Having had the opportunity to initiate, direct, and coordinate a number of projects related to research, curriculum development, university and community service, and manage change in an academic environment, I conclude one must first determine their present location in the aforementioned leadership quadrant, and determine where they want to shift and what level of responsibilities they want to have. After determining your quadrant position, identify two or three initiatives that are important to you in teaching, research, or service to the university. They may be your own, or you may support existing ones. These initiatives will be your *career drivers*. Now, it is your ability to grow as a leader that will determine your success and achievement in these initiatives. I contend that there are seven (7) traits and characteristics that must be evident to lead and accomplish goals with the support and cooperation of others. These are the academic qualities, characteristics, and practices that I believe are crucial to become an effective department chair. They are based on my past 10-year experience as a faculty member and 1-year experience as a department chair, and I am optimistic that they will have value in my effectiveness and survival in this new position. It is my belief that effective leaders become department chairs. Therefore, in my opinion, I recommend the following *leadership* characteristics to become, and perform as a department chair:

Listen, Observe, and Learn: One of the most important things you should do early in your career is to listen, observe, and learn. Just like every major corporation, every university also has its own culture. It is your responsibility to learn the informal and formal structure, personnel and relationships, academic politics, policies, practices, budgeting, and operations. It would be to your advantage to seek a mentor, and usually a senior faculty, dean, or seasoned department chair can take this role. However, this is not always the case, and you will have to become knowledgeable about the organization on your own. Learning how and what to do is an important skill, and knowing how to seek and obtain information is critical to becoming an effective leader. Also participate in internal and external faculty development activities such as workshops and conferences.

Communicate: This characteristic is common to all high achievers and leaders. The ability to communicate verbally and written is crucial to getting others to share your ideas and vision and become more visible. Become aware and learn all forms of informational technology resources available at your institution, and communicate with all levels of the organization. Practice and improve communication skills in teaching, research, and service opportunities. They will easily transfer to developing a strategy to pursue your career drivers within the university. Hence, share information with your dean and faculty constantly.

Be a Doer: Given the roles and responsibilities of faculty, become recognized as a person who gets the job done. Perform activities beyond the common faculty tasks. Learn to make your Chairman's job easier or even the Dean by completing departmental or college-level tasks. Demonstrate a commitment to follow up and accountability. These traits help establish a strong positive image and reputation.

Character: Personal traits such as honesty, integrity, and confidence determine who you are—your Image. Consistent with the values of the university, you should continue to develop these traits to reflect who you are, and what you stand for. Strive for excellence in performance and have respect for others. No one follows a leader with flawed or dissimilar character traits to accomplish anything.

Show Passion in Career Drivers: Everyone has his or her own personal agenda for certain things. Let your selected career drivers be your agenda and pursue with excitement and vigor. Leaders show enthusiasm in their agendas, and their agendas nourishe their soul. However, make sure your agenda doesn't conflict with more important initiatives as determined by individuals with more authority. Also be willing to support other agendas through partnerships and collaborations.

People Involvement: Though you may have your own career drivers, seek others who may share the same philosophy or interests at every level of the organization. Join selected committees with similar objectives as your career drivers. They can be your biggest supporters and help implement your initiatives. Not everyone will support your ideas, but make sure they understand them, and be willing to accept their perspective. Be willing to compromise and accept advice for the greater cause. They may be able to share and enhance your initiatives and make them more successful. This is important for a department chair.

Leadership Style: The biggest challenge for a leader is the ability to manage change, and people make change happen. Studies have shown that in any group an average of about 3% of the people are reformers, 16% adapters, 65% slow to change, and 16% resistors to change. Therefore, the style of leadership necessary to convert more than 80% of the people is critical to advancing your ideas. There are many styles of leadership. They vary from autocratic (e.g., I decide), democratic (e.g., We decide), empowerment (e.g., You decide), and everywhere in between. The key to successful and effective leadership contains the elements of fairness, ethics, competence, and decidedness. And most importantly, *others decide your leadership, not you*. Only then are they willing to follow. Learn from leaders in your organization, and take advantage of leadership seminars to develop a style that works for you and your ideas. Then develop a strategy to pursue your initiatives to improve the overall academic environment or workplace.

Listed above are self-nurturing activities and habits for faculty who would like to become stronger leaders and department chairs. Hence, I contend it will be these "leaders" who confront the challenges of higher education with creative solutions, and a motivation to serve. Find daily rituals to practice them while taking advantage of the opportunities and challenges that face higher education. It is only a matter of time before you see yourself moving toward the next higher quadrant.

To become an effective department chair, one should certainly understand and accept the gamut of responsibilities. Gmelch and Miskin (1995) provided a long list of responsibilities and tasks of a chair: study and learn them. They range from faculty recruitment and evaluation, departmental representation, to academic affairs coordinator and budget manager. However, little to no training is provided for department chairs in these and other areas (Creighton, 2001). There is the further challenge of managing the hectic schedule of teaching, research, and service. All of these become secondary as a department chair. Nevertheless, I would also recommend that as a new department chair, become familiar with university policies (faculty handbook), try to obtain some level of leadership training, and make sure you have the passion and personality for the position.

CONCLUSION

The objective of this paper was to share my perspective on leadership and give a retrospect on my first year experience as a department chair.

TABLE 9.1 Department Chair Evaluation Form
Faculty Evaluation of Department Chairman
Academic Year _____
Please rate the effectiveness of your Department Chair on each statement using the scale provided. Please check the appropriate box after each statement.

	1−Disagree 2−Needs improvements 3−Satisfactory 4−Exceed expectations	1	2	3	4
1.	Establishes clear goals and objectives for the department				
2.	Communicates specific departmental plans to accomplish goals and objectives				
3.	Implements departmental plans effectively				
4.	Clearly states expectations of faculty performance and duties				
5.	Encourages faculty discussion and participation in planning and decision-making				
6.	Demonstrates sensitivity to problems of departmental personnel				
7.	Makes equitable distribution of assignments to faculty members				
8.	Encourages good teaching practices and upholds the quality of teaching				
9.	Works toward updating curriculum				
10.	Encourages good teaching practices and facilitates professional development among faculty				
11.	Actively pursues attainment of grants, contracts, equipment, etc., from external sources				
12.	Gives feedback on individual faculty performance evaluation				
13.	Acknowledges and recognizes faculty contribution to departmental growth				
14.	Maintains confidence of faculty members				
15.	Effectively represents department to other administrative units				
16.	Pursues and maintains high academic standards				
17.	Is available and accessible to faculty and students				
18.	Demonstrates professionalism in interactions with faculty and students				
19.	Schedules an adequate number of departmental faculty meetings				
20.	Shows sound judgment in making decisions				
21.	Encourages cooperation among faculty				
22.	Demonstrates courtesy and respect for faculty, staff, and students				
23.	Serves as a positive role model for faculty				
24.	Overall, serves as an effective leader of the department				

This form is used exclusively to improve the management capability and effectiveness of your Department Chair. Your candidness and honest feedback is greatly appreciated. Thanks in advance. Department Chair.

Though many accomplishments were made during that year, I endured many obstacles and challenges as well. Some of the challenges are common in any leadership position in academia: faculty conflict and negotiation, budget management, limited resources, time management, curtailed research, and student issues (Hecht, 1998, 1999). As a department chair, these are compounded by the challenges of trying to maintain a certain level of research activity and the bureaucratic demands of university administration. Yet, this was the most exciting and rewarding year of my "academic career," and I look forward to continuing in this position for several years.

In this capacity, I plan to continuously evaluate my status and position on the Academic Leadership Quadrant, and monitor my transformation and growth of managerial skills and knowledge. I plan to continue my professional development through training and education, risky assignments, lofty goals, and a passion for succeeding while maintaining the values I believe in and support. I also encourage in this process, to seek evaluation of your leadership skills. Adapted from Gmelch and Miskin (2010), I developed a chair evaluation form for my faculty to assess my leadership of the department. It has proved useful in my own development and departmental management, which gave the faculty an opportunity to evaluate my performance (Table 9.1).

I hope the reader recognizes and deciphers that I believe leaders are *made* not *born*, and thus everyone has the capability to develop the character and commitment to make change and influence others toward a common cause. I challenge the reader to examine all available resources to support him/her in budgeting, planning, and so on in making this leadership position rewarding and beneficial to customers (Lucas, 2000). The best resource I have found is the ACE Department Chair Online Resource Center, and strongly encourage all chairpersons to regularly visit the website (www.acenet.edu). In engineering education, our students pursue a long journey toward understanding and learning, while initially professors provide the vehicular tools, leaders can provide the destination. I wish you luck and success in your call to leadership.

References

ABET.org, 2000. Accreditation Board for Engineering and Technology. Available from: <http://www.abet.org>.

Acenet.edu, 2002. The American Council on Education. Available from: <http://www. acenet.edu>.

Creighton, L., 2001. Running an Engineering Department Can Be One of the Toughest Jobs Around. PRISM, September, Vol. 11, No.1.

Cutlip, M., 2003. Departmental advisory boards—their creation, operation, and optimization. In: ASEE Annual Conference. June 22–23, Nashville, TN.

Elizandro, D., Matson, J., 2001. Industrial program management in the ABET 2000 environment. In: ASEE Annual Conference. June 24–27, Albuquerque, NM.

Fabiano, P., 2002. Strategic and operational planning at the department level. In: ASEE Conference, June 16–19, Montreal, Canada.

Felder, R., Brent, R., 2003. Designing and teaching courses to satisfy the ABET engineering criteria. J. Eng. Educ. 92, 7–25.

Gmelch, W., Miskin, V., 1993. Leadership Skills for Department Chairs, first ed. Anker Pub. Co., Bolton, MA.

Gmelch, W., Miskin, V., 1995. Chairing An Academic Department. Sage Publications.

Gmelch, W., Miskin, V., 2010. Department Chair Leadership Skills. Atwood Publishing.

Hargrove, S., Domingo, P., 2003. A Retention Study for the Industrial Engineering Department at Morgan State University. Internal Report, Morgan State University.

Hecht, I., 1998. The Department Chair as Academic Leader. ACE and Onyx Press.

Hecht, I., 1999. Transitions: from faculty member to department chair. Dep. Chair. 10, 2.

Kuo, W., Deuermeyer, B., 1998. IE curriculum revisted: developing a new undergraduate program at Texas A&M University. IIE Solut. 30, 16–22.

Leaming, D, 2003. Managing People: A Guide for Department Chairs and Deans. Anker Pubishing, Bolton, MA.

Leonard, M., 2003. Level reform of undergraduate industrial engineering education: a new paradigm for engineering curriculum renewal. In: ASEE Annual Conference. June 22–25. Nashville, TN

Lucas, A., 2000. Leading Academic Change: Essential Roles for Department Chairs. Jossey-Bass Publishing, San Francisco, CA.

McGinnis, L., 2002. A brave new education. IIE Solut. 34, 27–32.

Olds, B., Miller, R., 1998. An assessment matrix for evaluating engineering programs. J. Eng. Educ. 87, 173–178.

Shelnutt, J., Buch, K., 1996. Using total quality principles for strategic planning and curriculum revision. J. Eng. Educ. 85, 201–207.

Tucker, A., 1992. Chairing the Academic Department, second ed. American Council on Education, New York, NY.

Washington State University, 1992. Center for the Study of Department Chair, National Survey. Pullman, WA.

MENTORING

Mentoring

Rosemarie Tillman

Linn-Benton Community College, OR and Lane Community College, OR

In 2005, I conducted a qualitative study examining how faculty from two research extensive universities experience collegial behavior in the first 5 years of their appointment and developed a definition of collegiality from that data (Tillman, 2007). It was an interview study of higher education that identified collegiality through the advice participants would give to incoming faculty. My data was analyzed using thematic and sequential coding schemes. Five themes emerged: teaching, research, service, collaboration, and mentoring. In an effort to challenge the academy to reexamine initiatives for the diversification of American faculty, this chapter focuses on defining moments of collegiality as related to the mentoring theme of my 2005 study. I found mentorship to be important to faculty participants' definition of collegiality. Advice about mentoring covers two distinct types: formal and informal. Additionally I address mentoring within advice about research in terms of feedback and mentoring. I also discuss participants' personal notions of collegiality and the influence of its identification on mentorship. My hope is to illustrate some ways in which organizational processes, cultural stereotypes, and faculty mentorship can result in disadvantages for underrepresented faculty.

OBJECTIVE

The purpose of this chapter is to delve into dialogue and discussion about collegiality and mentorship. How is collegiality defined? How do faculty members behave collegially? How do faculty members experience collegiality? Why is collegiality significant? Does collegiality enrich faculty mentorship? Can the ambiguous aspects of collegiality adversely impact faculty mentees? Reanalyzing a study I previously conducted will hopefully answer these questions and provide some guidance along the varied paths to academic success and sustainability. There are many ways to mentor faculty members. My study's research question: "What advice do faculty at research extensive institutions give incoming faculty within the context of individual and organizational responsibilities?" revealed a variety of mentoring methods (effective, ineffective, positive, negative, insufficient, nonexistent) and a great deal of ambiguity within the understanding and practice of collegiality in higher education today.

BACKGROUND

Early in an academic career, collegiality can ease the way for new faculty by providing guidance about the "unwritten rules" that exist in the world of academia. During mid-career, collegiality can facilitate faculty transition from one institution to another. Late in a career, collegiality can be a sustaining influence on satisfaction as senior faculty are able to reach out and help others, using their own experiences and knowledge to aid newcomers (Miller and Noland, 2003). Faculty tutoring new faculty in the "unwritten rules" of a particular campus and a particular discipline can add positive relational dimensions to what is fundamentally a solo process (Smith, 2005). The existence of collegiality as a higher education norm that serves faculty supposedly lessens experiences of isolation and alienation and can increase each individual's ability to work successfully in academic settings. This belief is so widely held that collegiality is a taken-for-granted expectation of behavior on nearly every American campus (Tillman and Dunlap, 2007).

Yet, little research has been done on the phenomenon of collegiality. A few educational leadership and higher education scholars have written about the importance of collegiality in academic life and provided some starting point (Bess, 1988; Tierney and Bensimon, 1996; Kennedy, 1998; O'Brien, 1998). Bess (1988, pp. 104–110) presented collegiality as a multifaceted concept that has three faces: behavior, culture, and structure. Bess stated that collegiality is usually dealt with, thought about, and spoken of as a highly influential and concrete phenomenon; yet very few individuals stop to reflect upon this embedded normative phenomenon. Bess' line of reasoning begins discussing a variety of faculty role aspects, such as internal conceptualizations, role requirements, role expectations, and zones of ambiguity.

Tierney and Bensimon (1996) argued that erroneous beliefs about collegiality can seriously undermine, or perhaps even cripple, a scholarly career. They suggest that collegiality may partly consist in advice veterans give to new colleagues or it may consist of more intensive involvement in the new colleague's research or teaching activities. Veteran advice must draw upon individual and departmental experiences within collegial contexts and should shed light on what faculty believe about collegiality and collegiality's operation in their lives and their immediate professional environment.

Bess' and Tierney and Bensimon's studies acknowledged different facets of collegiality and the ambiguity of collegiality, but did not offer a way for faculty to distinguish one facet from another, or to develop a means for ensuring that discussions of collegiality are grounded in research, teaching, service, collaboration, or mentorship. However, when I searched the education literature for other work on prior definitions of collegiality, I found very little to assist me. Ultimately, communication theory helped me develop a more extensive theoretical framework for the study.

Communication theory explained the scripts and roles that govern collegial behaviors of faculty members. A script predicts behaviors and events (Trentholm and Jensen, 1992). There are scripts that tell faculty what should happen in various contexts or discussions about these same contexts. Scripts guide action by providing clues about how to proceed and what tasks to do in what order. Roles outline expectations (Trentholm and Jensen, 1992) and the behavioral patterns associated with members of a group (Adler and Rodman, 2009). The role of a faculty member (mentor, mentee, or colleague) dictates which behaviors to enact when as well as what thoughts to share and with whom.

In Smith's 2005 *Communication Education* article, clear scripts and roles are essential for socialization into a new academic community. Smith found that the most effective means of socialization is through mentorship. Smith's study asks for a reconceptualization of what it means to join and participate authentically and competently in an academic community (p. 69). Smith focuses on a "novice–expert discourse" in her discussions of improving interpersonal relationships, collaboration, negotiation, and pedagogy. The article explores transitions, norms, tensions, and ambiguities, and how faculty dyads manage to ebb or thrive together and individually under their influence. Smith's study shows collegiality as central to the communication processes of mentorship.

Collegiality is a process that most often occurs in the impersonal, public sector of social interaction where most of our relationships take place (Adler and Rodman, 2009). Any interaction is more difficult when you do not know the roles and scripts, or lines of action, but public interactions are virtually impossible without them (Trentholm and Jensen, 1992). This is complicated by the fact that collegiality is a highly complex and fluid process that differs from campus to campus, from discipline to discipline, and from one faculty cluster to another. People want to interact without feeling too uncertain, without investing too much of themselves, or without disrupting the perceived social order.

The complexity and fluidity of collegiality can make highly collegial and confident faculty uncertain when there is a shift in context. This uncertainty can become problematic when mentorship is a predominate context. This chapter reveals the conflict that is involved in mentorship and questions whether faculty mentorship provides a rich enough framework for understanding and managing collegiality as incoming faculty are inducted into an academic community.

METHOD

Interviews

For this study, it was important to establish trust in order to allow individuals to feel safe in reflecting upon and talking about a phenomenon that usually is not talked about: how collegiality was experienced, or not experienced, by each person. Because I wanted to single out constructed meaning by asking faculty to talk about their experiences of collegiality in different settings, my primary data collection method was an in-depth personal interview in a private setting. My methodological commitments were grounded in watching, listening, asking, recording, and examining the "everyday life world" of faculty members (Schwandt, 2001). I believe that meaning is constructed through thinking about and reflecting upon interpersonal relationships. These commitments and this belief encouraged me to ask questions such as:

> What construction informs what you think?
> What role are you playing?

Schwandt's (2001, p. 243) suggestion that "There are multiple, often conflicting constructions, and all (at least potentially) are meaningful" seems to apply directly to understanding a culturally embedded phenomenon such as collegiality.

Another formative methodological force in determining the research design of this study was my intent to let faculty responses and stories serve as my primary guide to construction of definitions of collegiality. For that reason, I decided to use open-ended questions and to use grounded theory analysis techniques. While I drew upon the theories of Bess, and Tierney and Bensimon, in particular, to frame the choice of open-ended questions, I was determined to pose questions that gave as little guidance as possible, so that I might have a broad response set.

I felt a broad response set might lead to the identification of embedded assumptions about collegiality (Glaser and Strauss, 1967).

Twelve faculty members were selected as a stratified purposive sample from a group of two Tier 1 research universities volunteers. The sample included nine senior faculty, and three junior faculty in the first 5 years of their first tenure track appointment. There were equal numbers of males and females, and equal representation from the hard sciences, the social sciences, and humanities departments. Of the two universities, one is private and the other public. Five faculty members were from the private institution and seven faculty members were from the public institution. While it was not the intent of this study to investigate possible "intervening variables" of gender, institution, or discipline, it was my hope that any identified anomalies in this structured sample might lead to recommendations for future research on the communication of collegiality within higher education. At the time, there was little prior research on this topic. Lack of research and because the phenomenon being explored was complex in nature (Borg and Gall, 2003) led me to chose this small strategic sample as the appropriate methodology.

Limitations were present in the nature of how the sample was constructed. Certainly the fact that I only interviewed people who volunteered to be interviewed biased the sample, but it is not possible to determine how that might be so. Using e-mail to solicit responses undoubtedly biased the sample to those who actively use e-mail. Time constraints limited both the data collected and the length of time spent with each interviewee. Results from such a limited sample and research conducted over such a limited time period must limit any attempts to generalize to larger populations based on these findings. However, generalizing in order to frame future research, to prior research, and to theory formation can be done.

Data Sources

Each faculty member was asked 18 questions about and around the topic of collegiality as drawn from higher education and communication theory. Interviews and field notes were transcribed and then blind coded for frequency and patterns by two coders, using the constructs of collegiality and communication theory.

RESULTS

Five themes emerged from the interview data: advice about teaching, advice about research, advice about service, the role(s) of collaboration, and the importance of mentorship communications. I have specifically focused this chapter on the advice given and received about the importance of mentorship. I synthesized the data about the importance of mentorship into three broad categories: feedback and mentoring in terms of research, formal mentoring, and informal mentoring.

Feedback and Mentoring in Terms of Research

The most consistent data set that emerged related to advice about what new faculty should do concerning research. This was not a terribly surprising finding, given that both institutions are research extensive universities and one might predict that advice about how to succeed with research would be central to any discussion of advice for success.

The advice that participants shared about research pertained to several forms of research related to publication: research experiments, research grants, research projects, research papers, refereed research journal articles, research books, and research-related presentations. I sorted 38 instances of research advice given or received by the 12 interviewees in three categories: advice on publications, advice related to the context of research publications, and individual feedback and coaching about the specifics of research. Almost all of the advice in the first two categories can be summarized in three simple statements. Do it. Do it a lot. Do it first.

The third category (individual feedback and coaching) shows how individuals actually gave each other feedback. For example, one faculty member (senior, male, hard sciences) talked four different times about advice on research, alluding to suggestions he had given. Once he referred to suggestions he received. Another faculty member gave a similar description, referring to this balancing act as a necessary "cost—benefit analysis" that each person must do to keep what is most important in the long run in front of each day's activities. Another stated it as "teach minimally." A third faculty member stated her advice as a "do not" as in "do not spend too much time on teaching," and "do not give too many guest lectures" so the focus can stay on publications. A fourth interviewee named "not publishing" as the critical error in succeeding made by a faculty member in his department and then completed his story by talking about how he coached other faculty members to help her. A female senior social sciences faculty shared instances where she sought out helpful hints. This same faculty member also shared about a piece of research advice in the form of feedback that was equivalent to a warning. A male senior hard sciences faculty spoke about an attempt to give research advice in the form of feedback that took a negative turn. The two remaining interviewees discussed talking with junior faculty about "what it takes to succeed at tenure." In fact, the specific advice given to someone who wanted to do something other than publish research was "leave."

Within respondents' discussion of feedback and mentoring in terms of research, Bess' three pivotal elements, or faces of collegiality (behavior, culture, structure), can clearly be seen. Collegiality as behavior is portrayed in that faculty were talking about the action of conducting research and the interactions that accordingly follow. These interactions are bounded by collegiality as structure (roles, rules, governance) and collegiality as culture (what's expected and accepted). The faculty interviewed for this study consistently named research publications as the foremost need for junior faculty to succeed. However, in their comments, they moved back and forth between Bess' categories of collegiality and did not specifically distinguish between a desired behavior, the specific context, and the particular rules of their institution. They merged all three together when they gave advice on feedback and mentoring in terms of research and when they talked about giving this type of advice.

At this point, Bess' definition of collegiality does not appear to be helpful in understanding how these faculty members viewed collegiality or communicated about it with other faculty. It seems evident that even very different faculty members in different departments and different institutions agree that research publications are critically important to getting tenure. All 12 interviewees commented on advice given concerning research publications, and that one important aspect of collegiality is giving advice about how to succeed with research publications.

Formal Mentoring

There were 20 instances of advice about mentoring that referenced formality. These instances relayed "official," "proper," and "recognized" advice about mentoring. Discussions included counseling from senior faculty, guidance through policies, direction for procedures, warnings about interactions, supporting scholarship, encouraging professional development, and advocacy with promotion and tenure issues. The definition of formalized mentoring that emerged from the participants' comments was departmentally or institutionally established relationships consisting of either one-to-one faculty, a mentoring committee of two to three senior faculty-to-one junior faculty, a variety of faculty advising a department chair, or some sort of corporate partnership.

Eight times committees or relationships formed under departmental direction were mentioned during discussions of advice about mentoring. Six respondents felt that departmental committees were worth mentioning. A male senior humanities faculty talked about teaching mentors. A male senior hard sciences faculty discussed a mentoring committee for the chair that other faculty sat on. A junior female hard sciences faculty shared about her research mentoring committee.

Three times university orientations were mentioned during discussions of formal advice about mentoring. Two respondents felt that these prescribed and often ceremonial activities were worth mentioning. Eight times department chairs were mentioned during discussions of formal advice about mentoring. Four senior respondents considered formal advice about mentoring from a department chair to be proper and something that should be standard. Three of these four respondents are currently department chairs.

Formal advice about mentoring was expected and was expected in various forms (e.g., collegiality as culture, or how it "should be"). Respondents felt that institutional stratification required organization at each level and imposes mentoring roles (collegiality as structure, or how "it is"). These participants argued that formal advice about mentoring should come from departmental committees, university orientations, and department chairs. Participants also talked about how important individuals were to them (collegiality as behavior, or how "it actually happens for me").

A female senior social sciences faculty also outlined another form of formal advice about mentoring. This additional form of formal mentoring involved corporate partnerships.

Informal Mentoring

This was, by far, the largest category of comments. Sixty comments on informal advice about mentoring were received from study participants, in addition to the 20 comments on formal advice. This would seem to indicate that communication of formal and informal advice to tenure track faculty was widely viewed as essential to success. Many participants stated that their mentoring situations were informal yet invaluable (collegiality as behavior, collegiality as structure, collegiality as culture).

Ten faculty members talked about not receiving any advice, receiving negative advice, and/or refusals to give advice, and clearly named this absence of informal mentoring as less than what should exist for new faculty. Five participants said that they received no advice about succeeding when they first came to their department. Three participants stated that they received little or minimal advice about succeeding. One participant said that it was his responsibility to give advice as an incoming senior faculty member. Another participant said "I always understood how universities worked and didn't need advice. I never got advice about anything." For this faculty member, his knowledge of collegiality as structure, the setup of collegiality, provided understanding of collegiality as behavior, actions, and interactions he should be involved with, and a general sense of collegiality as culture, or what is expected of a tenure tracked faculty member.

This was similar to Tierney and Bensimon's (1996) philosophy. They wrote about behavior codes that are routinely violated. For Tierney and Bensimon, passing comments about availability, or chance encounters that present opportunities or pressures for senior faculty to help junior faculty, are behavior codes that place faculty who do not know what questions to ask at a disadvantage.

Senior faculty can appear collegial by stating that they are willing to answer any questions a new faculty member may ask, but burdening new faculty with knowing what to ask and how to ask it can equal a refusal to help. One participant said that, in actuality, there are never any refusals to help. Another participant could not recall any refusals. A third participant never heard of refusals. However, three participants said that new assistant professors often have to know who to go to for help. For faculty to operate within the collegial structure of their environment, they must decipher unwritten norms of behavior. Figuring out how to enact norms involves finding out the expected norms. Expected norms are the link to Bess' collegiality as culture—how things should be. A male senior hard sciences faculty was a good example of this. "Uhm ... I did not receive much advice. I knew everyone and they probably assumed I knew the situations."

Most informal advice about mentoring was shared in an unprepared manner. Forty-six comments emphasized the relaxed nature of this type of advice. A female junior hard sciences faculty member spoke about important yet untailored advice from a senior member in her department. A female senior humanities faculty reflected on informal advice she received about the department chair when she first arrived. "She said, 'If he likes you you're golden, if he doesn't you're dead,' and that's for certain." She shared this after stating some informal advice she would give:

> Study your faculty, get the lay of the land, uh, and decide if you want to be in that department, decide if that's a place for you, if you'd be happy to work in, that it's a nourishing environment, if you have the skills to do what you would like to do, in terms of your own work, and in terms of teaching as well.

Whether it is formal or informal, advice about mentoring was plentiful and significant for faculty members, and to their academic lives. Regardless of male or female, senior or junior, discipline area, or public/private nature of the research extensive institution, all participants talked at length about formal and informal mentoring.

E. MENTORING

While all participants discussed advice about mentoring, most (but not all) participants talked about advice about research or teaching. Few participants mentioned service. No patterns were identified in any of the subcategories of advice related to the possible intervening "variables" of gender, time in rank, discipline, or institution. Participants who identified themselves as department chairs were more likely to talk about formal mentoring programs and activities.

PERSONAL NOTION

The three types of mentoring showcased in the data provide a look at how faculty members define collegiality. There are obvious times of confusion or uncertainty during each participant's discussion of the importance of mentorship. I asked the research question, *What advice do faculty at research extensive institutions give incoming faculty within the context of individual and organizational responsibilities?* Two perceptions from this data can be made. First, the largest number of advisory comments were made about mentoring. Therefore, mentoring in its many forms seems critical to any definition of collegiality. Second, although the number of comments about research were fewer than comments about mentoring, there was consensus that research and related publications outweigh all other factors in achieving tenure. Therefore, it would appear that directional advice about research and publications must be part of a definition of collegiality in a research extensive institution.

The times of confusion or uncertainty lead me to review participants' individual notions of collegiality. When I asked respondents to share their personal notion of collegiality in higher education, I wanted them to identify collegiality from their point of view. Participants were asked "How would you define collegiality?" in the hope of acquiring a sense of what comes to mind when they heard the word collegiality. I identified 21 quotes about what collegiality meant to these faculty members.

Many of the comments were similar to each other or combined similar aspects. Respondents felt that sharing of information, working together, and friendliness, along with being supportive, welcoming, and incorporating, were significant aspects of collegiality. Putting aside personal differences while looking out for other people, especially junior faculty, also had a significant showing in respondents' personal identifications of collegiality. These comments showed that, on a personal level, faculty members defined collegiality as an ongoing and sincere concern for the well-being and success of others that is overtly demonstrated on a regular basis.

The confusion and uncertainty noted in the data about the importance of mentorship can seemingly be attributed to a faculty participant's personal notion of collegiality changing based on whatever context the faculty member was describing at a particular time. This change in meaning did not seem to be an easy task for faculty. Participants appeared to mentally struggle with the changes they were reporting. At times, physical signs accompanied the mental signs of the struggle described. During interview sessions, some participants stared off for a few moments, held their head in their hands, scratched their head, used excessive verbal fillers ("uh," "hum," "ah"), took deep breaths, bit their lips, picked at their fingers, or grew red in the face. Respondents gave impressions of confidence about the communication of collegiality and initially seemed keenly aware of the communication of collegiality, yet their stories relayed not only discomfort and uncertainty, but often reflective tones of doubt about recent collegial interactions.

CONCLUSIONS

To summarize, faculty members define collegiality as an ongoing and sincere concern for the well-being and success of others that is overtly demonstrated on a regular basis. This definition emerged as respondents shared about the importance of faculty mentorship. In advice given and received about mentoring, a communication process surfaced: the communication of collegiality in higher education. It would seem that each participant came to terms with their individual notions of collegiality or experienced a defining moment of collegiality, within their attempts to discuss the importance of faculty mentorship.

I asked the research question, "What advice do faculty at research extensive institutions give incoming faculty within the context of individual and organizational responsibilities?" and the largest number of advisory comments were made about mentoring. Therefore, mentoring in its many forms seems critical to any definition of collegiality.

I did not expect the voluminous and conflicting advice given to new faculty about the importance of mentorship. Yet, upon reflection, I argue that the highly conflicting advice given to new faculty about mentoring clearly reflects the conflicts that sometimes arise in a research university between best practices of teaching and best practices for research.

Educational or Scientific Importance of the Study

My findings contribute to future education and communication research and also have practical importance for improving tenure track success, improving departmental collegiality, and improving overall educational quality. My findings also point to the embedded tension between teaching and research that is not always acknowledged by institutions and which may lessen individual chances for achieving tenure.

This study included data about collegiality in higher education that provides insight about faculty experiences of collegiality and potentially explains the significance and status of collegiality within academe. Clarification of the manifestations and mutations of collegiality can help us understand how collegial departments and campuses can attract (and retain) strong faculty, resulting in better teaching and/or academic productivity, and contributing to the quality of faculty work life.

The results of this study should encourage faculty to stop and think through interpersonal communications and interactions, attempt to determine specifically which meaning is dominant in an interaction, which level (interpersonal, departmental, institutional, or discipline) they are operating on, and then react appropriately. A wise saying from the Orient, found on Good Earth™ tea, renders it simplistically: "First think and then act."

Thinking through mentorship interactions becomes very important because, as William B. Johnson argues in his 2006 review of research on mentoring relationships, mentoring is crucial to the academy's survival. Johnson gives ample attention to the various types of mentorship found within higher education, consistently highlighting faculty benefits of successful mentoring. Additionally Johnson uses his psychology background to look at the keystones of negativity and find solutions when mentoring fails. Savage et al. (2004) seemingly agree with Johnson in their look at higher education's increasing substitution of new faculty orientation programs with mentorship programs. They also see induction of new faculty into the campus community as characteristic of good educational practice. Their discussion of the features and benefits of new faculty mentorship programs is laced with rationale and documentation of pitfalls of ineffective mentorship programs.

Problematic Trend Suggested by the Study

By focusing only on the mentoring theme of this study, I was able to see how faculty members define the process and practice of collegiality through faculty mentorship. As participants discussed their experience of collegiality, discrimination disguised as collegiality within mentoring relationships was often mentioned. It did not seem odd for faculty to experience, witness, or engage in discrimination under the pretext of collegial behavior. What I found to be peculiar was perplexity about whether collegiality as a tool of discrimination is insidious or incompetent.

Effective mentorship is vital to faculty success, and particularly important for success of underrepresented faculty. They conclude that women and minority faculty are less likely to be satisfied and successful even when engaged in mentoring relationships (Shollen et al., 2008). Their article discusses institutions not providing mentoring that successfully responds to the typical career challenges of unintended bias, the stress of biculturalism, work–family balance, cultural and social norms about acceptable behavior, and feelings of isolation.

It is common knowledge that discrimination is a serious problem on American campuses whether overt, covert, or ignorant. The goals of higher education are undermined and human dignity is offended where employment or education benefits, opportunities, or privileges are denied or restricted on the basis of race, religion, color, national origin, sexual orientation, disability, gender, or where a hostile environment exists (*The Bulletin of Buffalo State College*, 2008). Discrimination limits the opportunity for individuals to realize their potential and denies the rigors, joys, and fulfillment of intellectual curiosity.

This viewpoint is echoed in *The Sista' Network: African-American Women Faculty Successfully Negotiating the Road to Tenure* (Cooper, 2006). This is Terry Cooper's description of a qualitative study about covert faculty discrimination within institutional and departmental practices, and the subsequently damaging effects on African-American women and their careers. Cooper also encourages special types of networking within and across institutions, development and maintenance of confidence at all costs, and use of 12 principles for success and survival as an academic. Cooper found that for most African-American women faculty collegiality, most notable in terms of mentorship is one negative experience after another.

Regardless of concerns within the academic community, courts have affirmed the use of collegiality concerning faculty employment, promotion, tenure, and termination (Connell and Savage, 2001). With courts upholding collegiality as a relevant consideration of faculty evaluation, the ambiguity of collegiality as a concept, as well as the processes attached to this concept, make it imperative for faculty to be able to discern if the use of collegiality as a tool of discrimination is malevolent, so that they can determine action to ensure career success despite it.

The trouble is collegiality cannot be divorced from subjectivity and is always subject to abuse by dominant forces (Johnson, 2006). One reason collegiality cannot be divorced from subjectivity is it is rooted in "human communication [–] the process of creating meaning through symbolic interaction" (Adler and Rodman, 2009). The most significant feature of symbolic interaction is the arbitrary nature of symbols. "Symbols are used to represent things, processes, ideas or events in ways that make communication possible. ... We overcome the arbitrary nature of symbols by linguistic rules and customs. Effective communication depends on agreement among people about these rules" (Adler and Rodman).

Collegiality utilizes several types of symbolic interaction including intrapersonal, interpersonal, nonverbal, small group, and organizational communication. Examination of collegiality within faculty mentorship makes it easy to see the ambiguity of symbolic interaction with the continuous shifting between types. At times faculty involved in the same mentoring relationship do not make the shifts together, partly losing effective communication through simultaneous use of different rules. Discrimination can also hinder effective communication within mentorship and can be the cause of shifts between types of symbolic interaction.

Discrimination may not be intentional in that discriminatory actions are often reactions to deep-seeded beliefs or learned responses. There are faculty mentors who truthfully believe that they are behaving collegially and are unaware that their ability to notice subtle differences impacts how they interact with their faculty mentees. However, deliberate actions toward underrepresented faculty, by mentors who are fully cognizant of their prejudice and attempt to eliminate the characteristics of (or exercise judgment based on) subtle differences, are very purposeful and can subtly and adversely affect career success.

Discrimination can easily be disguised as collegiality because when both concepts turn into processes of practice they are usually based on personal opinions or feelings rather than on external facts or evidence. When a faculty member's collegiality is assessed, the concentration is on how an individual responds or behaves. Faculty may interpret their responses and behaviors to be perfectly acceptable, if these responses are not antagonistic, confrontational, or combative. Even still mentors who discriminate will assess similar mentees, with similar behaviors, who are members of people groups different from the dominant people group, differently. *In View of Academic Careers and Career-Making Scholars'* discussion concurs (Shaw, 2008). Vernon Shaw's opinion seems to be that academics need open minds when being collegial. Shaw sees the success of higher education being interdependent with individual faculty's career successes, yet no scholar's success is truly an individual endeavor. If decision-making scholars are closed minded, collegiality and productivity decrease as interactions become less meaningful and communication problems are more common. Closed minds cause barriers to successful integration into higher education.

Collegiality is important to academic success (Weeks, 2008). Discrimination can masquerade as collegiality often severely damaging if not all together destroying it. Knowledge gained by exploring this academic trend may increase understanding of collegiality, providing insight about faculty experiences of collegiality as a tool of discrimination within faculty mentorship, and potentially explicate the significance and status of collegiality within academe. Clarification of the manifestations and mutations of collegiality may shed light on discriminatory faculty mentorship, improving faculty communication, and increasing diversity of college and university faculty. This would promote the use of the educational benefits of diversity as an enriching resource for all who call the academy home for any length of time.

Recommendations for Future Research

Future research would be stronger if a study were designed to be a longitudinal-process study. The longitudinal-process approach recognizes that a [communication] system may profitably be explored as a continuing system with a past, a present, and a future; [taking] into account the history and future of a [communication] system and relating them to [its] present (Pettigrew, 1979). The communication of collegiality is a communication system.

Because of limited time, this study relied on descriptions of faculty's experiences to codify a point in time (the first 5 years of appointment). Another valuable study could examine not only how things were, but also how they are, and how they became that way. Such an examination would explore how collegiality is created, maintained, and dissolved.

Beliefs, power relationships, and cultural differences need to be studied in greater detail to understand how they support or undermine collegial behaviors. Collegiality and Christian faculty at secular institutions and collegial perceptions of gay, bisexual, and transgender faculty would be two studies that could potentially show support or undermining of collegial behaviors.

This study needs to be replicated with more faculty participants, with more variety in institutions types, and at different points in time, to understand if the theoretical operationalization of collegiality that emerged in this study is accurate for a broad range of faculty in higher education.

Further work needs to be done to relate the construct of collegiality to other constructs, such as collaboration, work satisfaction, and diversification of faculty. Studies exploring collegial behaviors of women faculty, navigation of academe by Black faculty, collegiality's intricacies for Asian faculty, and reflections of contingent faculty could all shed additional light on the communication of collegiality.

Further work needs to be done to more thoroughly relate the construct of collegiality to deeper theoretical streams, such as the work of Smith on the development of novice and expert relationships in hierarchical and domineering organizational structures and that of

Wolf-Wendel and Ward on faculty experiences fraught with difficulty and ambiguity.

Throughout this study, the senior respondents voiced concern that current legal and granting agency changes, combined with budgetary problems, were causing a reduction in collegiality. This certainly makes room for a study focusing on discussions department chairs have about the value of collegiality and the value of mentorship.

Administrators and the roles that they play were frequently mentioned by respondents. A study about how administrators foster and/or hinder mentorship could prove useful. I think it would be interesting to delve into the collegial dichotomy higher education administrators often confront. Writings on faculty incivility (e.g., elitism versus merit) may support investigation into the role collegiality plays in the lives of higher education administrators, especially as they develop education for existing faculty and socialization for new faculty (Twale and Deluca, 2008).

The nature of work within higher education continues to evolve, as higher education itself evolves, and as the larger society and the world evolve. Historically, we have valued the "community of scholars" so highly that the very notion of collegiality is so embedded in our ideal sense of higher education and the mentorship found there in that it is difficult to tell them apart. My findings ask for higher education to rethink what it means for faculty to proficiently join and truly participate in a community of practice and to rethink the ways in which faculty are mentored.

Faculty members would benefit from working together to create a shared understanding of what it means to engage in collegiality while teaching competently, without power-based interactions that unnecessarily confuse and frustrate. Is it not time for a deeper understanding of the concept of collegiality so that we can communicate and mentor in ways that enhance our teaching while protecting our collaborative practices and also extending them to the faculty of the future? When interactions are explored by means of collaboration and negotiation, the possibility exists for everyone's learning and success—incoming faculty, senior faculty, administrators, staff, and students—to be enhanced.

References

Adler, R.B., Rodman, G., 2009. tenth ed. Understanding Human Communication, vol. 561. Oxford University Press, New York, NY, pp. 96–110.

Bess, J.L., 1988. Collegiality and bureaucracy in the modern university: The influence of information and power on decision-making structures. first ed. Teachers College Press, New York, NY.

Borg, W., Gall, M., 2003. Educational Research: An Introduction. seventh ed. Allyn & Bacon, Boston, MA.

Connell, M.A., Savage, F.G., 2001. Does collegiality count? J. Coll. Univ. Law. Retrieved November 11, 2008 from American Association of University Professors, Washington, DC. Web site: < http://www.aaup.org/AAUP/pubres/academe/2001/ND/Feat/Conn.htm?PF=1 >.

Cooper, T., 2006. The Sista' Network: African-American Women Faculty Successfully Negotiating the Road to Tenure. Anker Publishing Company, Inc, MA.

Glaser, B.G., Strauss, A.L., 1967. The Discovery of Grounded Theory: Strategies for qualitative research. first ed. Aldine Pub. Co., Chicago, IL.

Johnson, W.B., 2006. On Being a Mentor: A guide for higher education faculty. first ed. Lawrence Erlbaum Associates, Mahwah, NJ.

Kennedy, D., 1998. Academic Duty. first ed. Cambridge, MA: Harvard University Press.

Miller, K., Noland, M., 2003. Unwritten roles for survival and success: senior faculty speak to junior faculty. Am. J. Health Educ. 34 (2), 84–89.

O'Brien, G.D., 1998. All the Essential Half-Truths About Higher Education. first ed. The University of Chicago Press, Chicago, IL.

Pettigrew, A.M., 1979. On studying organizational cultures. Adm. Sci. Q. 24, 570–581.

Savage, H., Karp, R.S., Logue, R., 2004. Faculty mentorship at colleges and universities. Coll. Teach. 52, 21–24.

Schwandt, T.A., 2001. Dictionary of Qualitative Inquiry. second ed. Sage Publications, Thousand Oaks, CA.

Shaw, V.N., 2008. In View of Academic Careers and Career-Making Scholars. first ed. Information Age Pub, Inc. Charlotte, NC.

Shollen, S.L., Bland, C.J., Taylor, A.L., Weber-Main, A.M., Mulcahy, P.A., 2008. Establishing effective mentoring relationships for faculty, especially across gender and ethnicity. Am. Acad. 4, 131–158.

Smith, E.R., 2005. Learning to talk like a teacher: participation and negotiation in co-planning discourse. Commun. Educ. 54 (1), 52–71.

Tierney, W.G., Bensimon, E.M., 1996. Promotion and Tenure: Community and socialization in academe. first ed. State University of New York Press, Albany, NY.

Tillman, R., 2007. The communication of collegiality: An examination of the advice faculty give incoming faculty. (Doctoral dissertation, University of Oregon, 2006). Dissertation Abstracts International (UMI No. 323876).

Tillman, R., Dunlap, D., 2007. Definitions of collegiality in today's research universities. A paper presented at the annual meetings of the American Educational Research Association, Chicago.

Trentholm, S., Jensen, A., 1992. Interpersonal Communication. second ed. Wadsworth Publishing Company, Inc. Belmont, CA.

Twale, D.J., Deluca, B.M., 2008. Faculty Incivility: The rise of the academic bully culture and what to do about it. first ed. Jossey-Bass, San Francisco, CA.

Weeks, K.M., 2008. Collegiality and the quarrelsome professor. Lex Collegii: A Newsletter for Higher Education, Retrieved November 7, 2008 from College Legal Information, Inc., Nashville, TN. <http://www.coleggelegal.com/lccolleg.htm>.

Further Reading

Baron, J.L., 1984. Organizational perspectives on stratification. Ann. Rev. Sociol. 10, 37–69.

Bellas, M.L., 1999. Emotional labor in academia: the case of professors. Emotional labor in the service economy. Ann. Am. Acad. Polit. Soc. Sci. 561, 96–110.

Gall, M.D., Gall, J.P., Borg, W.R., 2003. Educational Research: An Introduction. seventh ed. Allyn & Bacon, Boston.

Guralnik, D.B. (Ed.), 1974. Webster's New World Dictionary of the American Language. William Collins + World Publishing, Cleveland, OH.

Keltner, J.W., 1970. Interpersonal Speech-Communication: Elements and Structures. Wadsworth Publishing Company, Inc, Belmont, CA.

O'Hair, D., Stewart, R., Rubenstein, H., 2007. A speaker's Guidebook. third ed Bedford/ St. Martin's., Boston.

Sanders, I.T., 1973. The university as community. In: Perkins, J.A. (Ed.), The University as an Organization. McGraw Hill Book Company, New York, pp. 57–78.

Schuster, J., Finkelstein, M., 2006. The American Faculty: the Restructuring of Academic Work and Careers. Johns Hopkins University Press, Baltimore.

Simpson, D.J., Thomas, C., 1993. Community, collegiality, and diversity: Professors, priorities, and perversity. In: Van Patten, J.J. (Ed.), Understanding the Many Faces of the Culture of Higher Education. The Edwin Mellen Press, Lewiston, pp. 17–31.

African-American Researchers in Computing Sciences: Expanding the Pool of Participation

LaVar J. Charleston[1], Jerlando F. L. Jackson[2], Juan E. Gilbert[3] and Ryan P. Adserias[1]

[1]Wisconsin's Equity and Inclusion Laboratory (Wei LAB), University of Wisconsin-Madison, WI [2]Department of Educational Leadership and Policy Analysis & Wisconsin's Equity and Inclusion Laboratory (Wei LAB), University of Wisconsin-Madison, WI [3]Computer & Information Science & Engineering Department, University of Florida, FL

The National Science Foundation (NSF) and the National Science Board (NSB) have each issued pressing warnings about the loss of US dominance in critical areas of science and innovation and the shortage of US scientists entering technical fields (CACM News Track, 2004). Additionally, Department of Labor projections show information technology (IT) job growth outstripping IT degree production for the foreseeable future. For decades, this problem has been on the horizon and research has pointed toward a deceptively simple solution: increase US citizens' participation in science, technology, engineering, and mathematics (STEM) disciplines, particularly among under-represented populations. However, broadening participation in the STEM workforce remains an ongoing challenge. The potentially lucrative careers of STEM professionals would seem to be a major draw for joining the STEM workforce, yet the promise of high earning jobs alone is not enough to overcome the many barriers to careers in STEM fields

Navigating Academia: A Guide for Women and Minority STEM Faculty.
DOI: http://dx.doi.org/10.1016/B978-0-12-801984-9.00011-0.

experienced by underrepresented populations—African-Americans especially.

This chapter explores the known reasons why African-Americans do not pursue or persist in STEM disciplines in general and in computing sciences in particular.[1] Likewise, this chapter presents data showing low numbers of African-Americans within computing science faculty ranks. The African American Researchers in Computing Sciences (AARCS) model will be examined and presented as an intervention aimed at increasing the number of African-Americans pursuing doctoral degrees in computing sciences. The ultimate goal of AARCS is to increase the number of African-Americans obtaining an academic faculty or research scientist position. Empirical findings from a study will be presented that demonstrate the AARCS model's ability to create behavioral, affective, and cognitive change in African-American undergraduate students in the computing sciences.

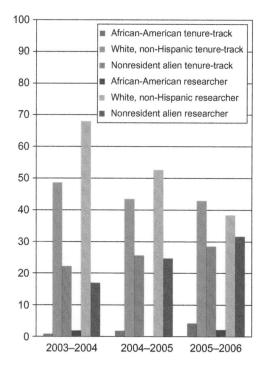

FIGURE 11.1   2003−2006 Ethnicity of newly hired faculty (tenure-track/researcher).

[1]In this chapter, the term computing sciences is used to refer to computer science and computer engineering.

AFRICAN-AMERICANS IN COMPUTING SCIENCES

According to the Computing Research Association (CRA) Taulbee Survey, African-Americans represent just 1.3% of all computing sciences faculty (CRA, 2006) (Figure 11.1). Nationally across all disciplines, African-Americans represent 5.2% of all academic faculty (Harvey and Anderson, 2005). The AARCS program takes steps to narrow the gap between computing science faculty and the national average by eliminating concerns, misunderstandings, and misinformation about graduate school, research, and computing sciences faculty among African-American undergraduate computing sciences majors. According to the Taulbee Survey, 460 (4.2%) and 18 (1.3%) African-Americans obtained bachelor and doctoral degrees in computing sciences, respectively, in 2005–2006 (CRA, 2006). In contrast, White and non-Hispanics obtained 5,805 (63.7%) and 376 (26.2%) of the bachelor and doctoral degrees in computing sciences, respectively. Also, nonresident aliens obtained 794 (8.7%) and 814 (56.8%) of the bachelor and doctoral degrees in computing sciences. The ratio between bachelor degrees awarded to doctoral degrees awarded for African-Americans, White, non-Hispanics, and nonresident aliens is 4:1, 7:1, and 100:1. The gap between the three groups clearly shows room for improvement, especially for African-Americans.

With respect to faculty hiring, just two African-Americans obtained tenure-track faculty positions in 2004–2005 (CRA, 2005), or just 0.8% of all tenure-track hires in computing science (see Figure 11.1). Compared to White, non-Hispanic, and nonresident alien populations, African-Americans are severely underrepresented at the tenure-track faculty ranks in computing sciences. Such disparities are not surprising given the presence of so few African-Americans in doctoral programs in the field. Since 2000, African-Americans have consistently represented no more than 2% of students enrolled in or graduating from computing sciences PhD programs. Given African-American representation in computing sciences PhD programs and graduation rates of 2% or less, low numbers of faculty positions obtained by African-Americans have followed suit.

In 2006, African-Americans received 4.3% of the tenure-track faculty positions in computing sciences (CRA, 2006) (see Figure 11.1). This milestone increase in the share of African-American faculty in computing science corresponded to the inaugural year of the AARCS program.

The disconnect between the number of bachelor degrees earned by African-Americans and the number of doctoral degrees awarded warrants further investigation and intervention. The disparity in the number of African-Americans at all faculty levels in computing science departments at our nation's colleges and universities is particularly troublesome when considering the large number of tenure-track positions held by White, non-Hispanics, and nonresident aliens.

While the AARCS is neither the first nor only intervention aimed at increasing the number of minority faculty in STEM field, it is unique in its approach. Having established successful programs such as the Meyerhoff Scholars Program (MSP) (Gordon and Bridglall, 2004) and the Bowling Green State University Academic Investment in Math and Science (BGSU AIMS) (Gilmer, 2007) devote considerable attention to the academic development of program participants. AARCS, by contrast, focuses mainly on providing information, exposing students to mentors and role models, as well as opportunities to build a network of computing science resources. The AARCS program also differs from other STEM-focused interventions in its prioritizing of computing science over other STEM fields. While the MSP and BGSU AIMS programs certainly support and help develop aspiring computing scientists, the AARCS program is singularly tailored to the experiences of African-Americans in computing sciences.

MOTIVATIONS FOR ESTABLISHING THE PROGRAM

In 2004 and 2005, a male African-American computing sciences faculty member accompanied by several other African-American graduate students traveled to Spelman College in Atlanta, Georgia, to deliver a presentation about barriers to computing science education (discussed in more detail below). The purpose of these trips to Spelman and the graduate school presentation was to recruit Spelman graduates to apply to enter the presenter's graduate program. After the 2005 visit, the presenter's computing sciences department received six applications from Spelman graduates, up from just the single Spelman applicant the program typically saw in previous years. The increase in applications was notable, as the six applicants constitute one-third of Spelman's 18 computer science seniors. Although unproven (albeit grounded in strong anecdotal evidence), it was hypothesized that the presentation was likely responsible for the spike in Spelman applicants.

This success story became the impetus behind the AARCS program and an effort to formally evaluate whether the presentations, identified by the AARCS program as a "targeted presentation (TP)" that addresses the seven identified barriers or pathways to persistence, significantly influence African-American undergraduates to apply to computing science graduate programs. Evidence from the AARCS study shows that indeed, TPs do have its intended impact on African-American undergraduate computing science students.

THE AARCS PROGRAM

The AARCS program consists of three components: targeted presentations, future faculty mentoring, and an annual AARCS mini-conference.

Targeted Presentations

The TPs component was derived from the 2005 Spelman College presentation. In this component, at least one faculty member and one graduate student travel to a historically Black college or university (HBCU) and/or a predominantly White institution (PWI) to give a presentation discussing graduate school, computing sciences research, and academic faculty employment. These presentations have been designed to address seven barriers identified as obstacles to matriculation and persistence in STEM fields. The faculty member is the chief facilitator of the TP with the graduate student assisting in answering questions following the presentation. The TPs address common misunderstandings undergraduates may have about computing sciences graduate school programs, academic faculty, and research. The content of the TPs address each of the seven barriers as follows:

1. Stereotypes
   Students typically harbor negative stereotypes of scientists, which are immediately broken down when the faculty member and graduate students walk into the room. None of those stereotypes fit the presenting faculty or the graduate students. In the presentation, the stereotypes are pointed out and immediately discarded.
2. Role models
   The presentation explains the importance of role models. At the same time, the presentation reveals the low numbers of African-Americans in the computing sciences. This approach is taken to help the students realize the slim chances of finding African-American role models within the computing sciences. However, through AARCS, the students will gain entry to a network of African-Americans in computing sciences.
3. Helping professions
   Within the presentation, several links are made to illustrate how computing sciences can be used to "give back" and help others. Specifically, the presentation describes research areas (e.g., artificial intelligence, advanced learning technologies, and human-centered computing) that serve as vehicles for giving back, demonstrating how computing can be used to help others.

4. Financial concerns

The presentation covers graduate school funding opportunities including an overview of how graduate teaching assistantships, research assistantships, and fellowships may be available to help defray the costs of graduate education. Specific fellowship opportunities for African-Americans are also discussed.

5. Inadequate advisement

The TP and the AARCS program are vehicles for proper advisement; therefore, the presentation itself addresses this issue.

6. Lack of knowledge regarding the advantages of having a PhD

The presentation presents several advantages of obtaining a PhD in computing sciences (e.g., tenure, the ability to work on problems of personal interests, salary).

7. Employment opportunities

Employment opportunities are addressed by providing facts about computing sciences. The presentation addresses outsourcing concerns, corporate employment options, government research opportunities (e.g., NSF and DARPA), faculty employment options, and research scientist options.

The TPs fill critical information gaps among undergraduates who may otherwise fail to consider pursuing graduate school opportunities. Additionally, once enrolled, students are provided further motivation to persist in their studies through the future faculty mentoring component.

Future Faculty Mentoring

Given the disparity between the number of African-Americans receiving PhDs in computing sciences and the number who pursue faculty positions, it became clear that African-Americans do not receive adequate advice upon the completion of their doctorate. As a result, the Future Faculty Mentoring Program (FFMP) was developed as a second component of the AARCS model. The FFMP aims to address the concerns of graduate students and recent graduates who have matriculated through computing sciences graduate programs but may not feel adequately prepared for careers as faculty or research scientists.

Initially, the FFMP drew upon a group of African-American faculty from across the country in computing sciences PhD programs at research intensive institutions. The goal of the FFMP component was to advise advanced graduate students on the academic job search process. Each of the participating graduate students self-identified an interest in obtaining a faculty position, but no one had explained to them how to search for positions, interview effectively, negotiate salary, etc. In fact, some of the students had been tacitly discouraged from pursuing research faculty positions and told instead that they would make "good teachers."

Mentoring is a critical component in the professional and personal development of graduate students, and particularly for future faculty. Jacobi (1991) described three features necessary in an effective mentoring relationship: emotional and psychological support, assistance in the career and professional development of mentees, and role modeling. All three aspects have been interwoven throughout every aspect of the FFMP in the AARCS.

The FFMP for advanced graduate students is grounded in empirical research that highlights the role of mentoring in graduate students' professional development (Hall and Burns, 2009). Research cites mentoring as an effective component in the retention of underrepresented science students, such as women, in computing science programs (Cohoon, 2001). In her study of female undergraduate students in computing sciences, Cohoon (2001) demonstrates that increases in mentoring helps moderate attrition rates among women. Cuny and Aspray (2001) of the CRA recommend making mentoring of women computing science graduate students a departmental priority. As members of an underrepresented population in the sciences, women are shown to respond particularly strongly to mentoring opportunities (Bird and Didion, 1992).

Other findings show same-race mentoring plays a positive role in the personal development of African-Americans at PWIs (Patton and Harper, 2003). Patton and Harper (2003) demonstrate how mentoring can impact the success of African-Americans in predominantly White spaces. For the African-American women interviewed in their 2003 study, Patton and Harper (2003, p. 74) found evidence that similarly identified mentors contributed to a more dynamic and fulfilling academic experience and are "vitally important." Essien (1997) shows how the limited number of similarly identified faculty in graduate departments may cause African-American women to seek mentorship outside of their field or within informal networks. The FFMP component of the AARCS program provides participants with the opportunity to establish mentoring relationships among African-American scholars and professionals in the field of computing science and to establish ties that aid in developing a robust network of professional contacts.

The FFMP group exchanges e-mail on a regular basis and participants engage in regularly scheduled conference calls. Some of the FFMP activities include reviewing academic job announcements and offers collected from other African-Americans in computing sciences. The students share information about their interviews, prospects, and experiences. In fact, the FFMP offers participants access to a database of job offers that can be used for future FFMP groups. The success of the FFMP is generally measured by FFMP participants' acceptance of a faculty or research scientist position.

AARCS Mini-Conference

The AARCS mini-conference is a 2-day symposium that brings together students from the TP sites with African-American researchers and faculty in computing sciences all over the nation. Undergraduate students apply for travel scholarships to attend the AARCS mini-conference. The mini-conference includes the following activities:

1. Prominent African-American researchers from computing sciences serve as speakers.
2. Each student participant is assigned a faculty and graduate student mentor based upon their research interest. These groups meet at AARCS to establish mentoring relationships and to discuss research, graduate school, etc.
3. The FFMP group also attends the AARCS mini-conference. Senior African-American faculty and former FFMP group members offer sessions targeted to the FFMP group.
4. Research writing and presentation sessions are conducted.
5. A TP is given.
6. Grant writing workshops are conducted.

Evidence of Success: Descriptive Summary

Targeted Presentation

Quantitative data. From 2006 to 2009, 232 students attended the TPs. The vast majority of participants were African-American/Black (95.7%) with the remaining identifying as Hispanic (0.4%), Asian/Pacific Islander (1.7%), White (1.7%), or American Indian/Alaskan Native (0.4%). Just over half were male (55.5%). A plurality came from low-income, single head of households and middle-income households. The majority (60%) of the TP participants was involved in extracurricular activities emphasizing computing sciences or undergraduate research programs and likewise, had very high levels of contact with faculty by all measures. Furthermore, after attending the presentation, approximately 70.7% (compared to 50.0% prior) were likely or very likely to state an intention to attend graduate study in computing sciences. Additionally, after attending the presentation, 81.5% were likely or very likely to submit an application for graduate school in 1–5 years in any field.

Post-presentation, 69.2% of the participants (compared to 12.8% prior) changed their views for the better regarding scientists and computing sciences as a career option. Hence, approximately 87.0% felt that graduate school was a possible career path for them after attending the presentation. Finally, approximately 51.8% of the TP participants could see themselves as faculty or researchers in computing sciences after attending the presentation.

Qualitative data. Written survey responses were collected from TP participants and were used for both program development and research purposes. Written feedback from the TP is presented here to supplement the quantitative survey data collected.

Responses from the participants illustrate the efficacy of the TPs in addressing and dispelling the misinformation, stereotypes, and misunderstandings of African-Americans in computing sciences. One participant of a 2008 TP wrote: "It's always nice to have the lack of African-Americans and women to be put into perspective [and] to help motivate soon-to-be college graduates." Another participant of the same TP cited their experience as, "[Clearing up] a lot of stereotypes… about going into Academia," while a 2006 TP participant reported that it "resurged my drive to pursue a PhD." Finally, one participant cited the TP as having "sparked my interest in possibly teaching in computer science field, (*sic*) whereas before I was only considering working in the corporate field."

The qualitative data show participants in TPs came away from their experience having their eyes opened to the possibilities of computing sciences, not only for African-Americans generally, but personally. More encouragingly, data from the TP evaluation responses show TPs effectively address six of the seven AARCS-identified barriers to persistence in computing sciences. Except for the perception that African-Americans are best suited for, or more attracted to "helping professions," participants cited TPs as allaying their fears concerning issues of stereotypes of computing scientists, the lack of African-American role models, concerns about financing graduate education, inadequate advising, and the lack of knowledge regarding the advantages of having a PhD or the career prospects for computing scientists.

AARCS Annual Conference

Quantitative data. There were 120 participants who attended the AARCS annual conferences (ACs). Almost all attendees were African-American/Black (95.8%) with the remaining identifying as Hispanic/Latino (0.8%), Asian/Pacific Islander (1.7%), White (0.8%), or American Indian/Alaskan Native (0.8%). The gender balance was tilted toward females (58.3%), while a plurality were from middle-income households. The AC participants had significant experiences that traditionally lead to research-related positions in computing sciences. Furthermore, these AC participants had a very high level of contact with faculty by all measures. Among the AC participants, 99.2% of them felt the conference proved informative or very informative. Likewise, 100% of the AC participants felt the conference was successful in connecting them with role models in computing sciences.

Qualitative data. Twenty-six African-American attendees participated in focus groups (73.0% male, 26.0% female). All of the interviewees had either majored in or were majoring in an area related to computing sciences as undergraduates. Each of the interviewees indicated that their first experiences within computing sciences occurred in middle or high school through coursework in computing. Three overarching themes emerged from the focus groups concerning participants' exposure to, and initial interest in, computing sciences: (1) parental purchase of computers at an early age; (2) classroom exposure to computer programming; and (3) an incremental interest in mathematics progressing to computer science. No one interviewee was able to completely separate their interests in the computing sciences from other noncomputer sciences-related interests. Videotaped focus group data was collected on site at the annual AARCS mini-conference from self-selected participants, who subsequently provided verbal consent for their participation. Interviews lasted between 10 and 40 min and were transcribed for data analysis.

With regard to the first theme of home computer adoption, one 2008 focus group member recalled, "When I was 8, my grandma bought a Tandy. I stumbled upon Basic programming language." Students who reported introduction to computer programming in the classroom also cite the experience as formational. One graduate student focus group participant was able to recall a high school teacher who was particularly knowledgeable and contributed to the student taking an interest in computing science:

> I went to a poor high school at first, and then I went to a private high school. And they had some programming classes. You know, just Basic and a few other little things. I guess that was my first experience [with computer programming]. My teacher was Mr. Bethel. He really knew his stuff.

The third and final major theme to evolve from the focus groups was the role of related coursework in mathematics. One participant reported switching majors from mathematics to computer science on the recommendation of an advisor: "My math advisor told me computers were the thing of the future, so I switched my major from math to computer science."

Future Faculty Mentoring Program

Quantitative data. There were 21 participants involved in the FFMP. The vast majority were African-American/Black (78.6%) with slightly higher proportions of White (7.1%) and Asian/Pacific Islander (14.3%) participants compared to the other components. Females were also represented in higher numbers (71.4%). The largest share of participants came from middle-income households. FFMP participants also reported a very high level of contact with faculty by all measures. Among the FFMP participants, 71.4% had been involved in an undergraduate research program. Likewise, 71.4% of the FFMP participants felt their faculty advisor was a mentor to them. The majority of FFMP participants had faculty advisors of the same gender, but a different race. Most of the FFMP participants (85.7%) felt that participation in FFMP made them feel optimistic or very optimistic about the job search process. Likewise, 92.9% of the FFMP participants felt the program provided them with concrete information with regard to pursuing faculty and research positions.

Qualitative data. Similar to participants of the targeted programs, FFMP participants were asked to provide written feedback to aid in both understanding program efficacy and to help guide future iterations of the program. This qualitative data was collected along with the survey results and analyzed to gauge the strengths of the FFMP as it relates to prior research concerning mentoring relationships for graduate students. In general, participants' responses to the FFMP were positive, and in line with prior research on the efficacy and importance of mentoring relationships between established faculty and graduate students. For example, one student credited the FFMP with helping to secure faculty interviews, "It provided specific contacts that 'fast-tracked' my application. Many of my peers had no responses from schools. However, I was invited to give 3 talks, 2 as a result of program contacts." This result aligns with the career development component of the foundational aspects of mentoring relationships, as outlined by Jacobi (1991).

Another FFMP participant from the 2008 cohort described the program's psychological benefits. While the process of traversing the job market may be frustrating at times, this participant wrote, "[The FFMP was] very positive and encouraging. I was very pessimistic upon initially applying to positions. I almost applied to no schools at all." Further evidence coincides with Patton and Harper's (2003) study showing access to African-American mentors to be beneficial. One graduate student cited the FFMP as, "very helpful for students who have been isolated on predominantly White campuses and departments." While no participants cited the FFMP as influencing their decision to persist toward their doctoral degree, several participants credited the program with increasing their confidence in establishing careers as African-American computer scientists.

CONCLUSION AND IMPLICATIONS

This book chapter provides empirical evidence of a successful intervention aimed at increasing the likelihood that African-Americans pursue graduate education in the computing sciences. Results suggest that the components of the AARCS program positively changed the attitudes and perceptions of participants about the field, graduate school aspirations, and their outlook on research scientists. While this intervention was ultimately aimed at adjusting perceptions and attitudes—only longitudinal data collection and analysis can speak to behavioral outcomes—the pre- and posttest features of the AARCS program empirically demonstrate the program's promise in improving computing science persistence through the highest degree attainment levels among African-American students. As such, the three components of AARCS (i.e., TP, FFMP, and AARCS mini-conference) indeed achieved its objective to empirically show behavioral, affective, and cognitive changes in African-American computing sciences aspirants. AARCS-derived quantitative and qualitative data demonstrates newly found interest in computing sciences doctoral studies, academic faculty positions, and researcher positions among the program participants.

By addressing the seven barriers to computing science persistence through advanced degree attainment, the TPs of the AARCS program facilitated a 21% increase in the likelihood that participants would state an interest in attending graduate study in the computing sciences. Likewise, about 82% of TP participants indicated their intent to submit an application for graduate school in 1–5 years in any field. These data indicate that the TP indeed had a positive effect on participants by exposing them to the attainability and benefits of graduate education. Additionally, there was a 57% increase in positive views of scientists and computer science careers post-presentation. Likewise, about 52% of participants reported that they could see themselves as faculty or researchers in computing sciences after attending the presentation. These data vividly indicate that an intervention such as AARCS may be a positive first step in increasing faculty and researcher participation in the computing sciences among African-Americans.

As the computing sciences can be a very isolated field, particularly for the underrepresented African-American, the AARCS mini-conference also provided a mechanism for participants to: (a) socialize with other African-American computing aspirants; (b) pair with other researchers of similar interest; (c) gain information about technical aspects of computing; and (d) gain valuable information about computing science careers as well as job opportunities. This assertion is buttressed by the 99.2% of study participants who reported that the conference proved informative or very informative. Additionally, AC participants unanimously felt the conference was successful in connecting them with role models in computing sciences. As such, these data reveal the necessity of socialization to the field of computing; particularly among African-American computing aspirants and

potential contributors to the field. Because these data indicate participants felt motivated to persist after attending the conference, this
aspect of the intervention proved essential to efforts to increase
participation—particularly among the graduate and doctoral ranks
within computing sciences.

Finally, the FFMP component of AARCS provided participants
greater confidence in their pursuit of a faculty or research position. By
and large, FFMP participants felt that participation in the mentoring
program made them feel optimistic or very optimistic about the job
search process (85.7%). Likewise, almost all (92.9%) of the FFMP participants felt the program provided them with concrete information with
regard to pursuing faculty and research positions. This aspect of the
intervention aided participants in their preparation for job attainment,
thereby increasing the likelihood of broadening the participation of
African-Americans among the faculty and researcher ranks in the computing sciences.

Ultimately, AARCS is intended to serve as a model to spark and nur-
ture the interests of African-Americans toward the computing sciences
field. However, as a model for broadening participation, lessons drawn
from this evaluation have implications for current and future programs,
which must first address the misconceptions, attitudes, and misunderstandings that endure among underrepresented groups about
computing sciences. By addressing these mental barriers, government
programs, college, and even K–12 curricula can dispel damaging myths
about computing sciences, which may otherwise interfere with the
career paths of underrepresented populations—namely, African-
Americans. As such, current faculty and researchers within computing
sciences, especially African-Americans, must also assist in addressing
these barriers to potential contributors to the field. This can be done
through grants, mentorship, as well as outreach and teaching.

As the demographic landscape continues to change in the United
States, it is becoming increasingly urgent that participation in computing be broadened in order for the United States to remain globally
competitive and keep up with labor market demands. As African-
Americans remain a virtually untapped pool of potential applicants in
the computing sciences, the AARCS model serves not only as an example of a strategy for increasing African-American participation, but also
for increasing participation of all underrepresented groups. The expan-
sion of programs like AARCS described and assessed in this book chapter could prove beneficial in broadening participation generally. If
effectively applied, such efforts can ultimately bridge the gap between
the jobs that need to be filled in computing, and an increasingly diverse
American workforce seeking to make a contribution.

References

Bird, S., Didion, C., 1992. Retaining women science students: a mentoring project of the association for women in science. Initiatives. 55, 3–12.

CACM News Track-Dominance Lost, 2004. Commun. ACM. 47, 9.

Cohoon, J., 2001. Toward improving female retention in the computer science major. Commun. ACM. 44, 108–114.

Computing Research Association, 2005. CRA Taulbee Survey. Available from: <http://www.cra.org/statistics/>.

Computing Research Association, 2006. CRA Taulbee Survery. Available from: <http://www.cra.org/statistics/>.

Cuny, J., Aspray, W., 2001. Recruitment and Retention of women graduate students in computer science and engineering. first ed. Computing Research Association, Washington, DC.

Essien, F., 1997. Black women in the sciences: challenges along the pipeline and in the academy. In: Benjamin, L. (Ed.), Black Women in the Academy: Promises and Perils, first ed. University of Florida Press, Gainesville, FL.

Gilmer, T., 2007. An Understanding of the Improved Grades, Retention and Graduation Rates of STEM Majors at the Academic Investment in Math and Science (AIMS) Program of Bowling Green State University (BGSU). J. STEM Educ. 8.

Gordon, E., Bridglall, B. 2004. Creating excellence and increasing ethnic minority leadership in science, engineering, mathematics and technology: a study of the Meyerhoff Scholars Program at the University of Maryland, Baltimore County (Unpublished).

Hall, L., Burns, L., 2009. Identity development and mentoring in doctoral education. Harv. Educ. Rev. 79, 49–70.

Harvey, B., Anderson, E., 2005. Minorities in Higher Education Twenty-First Annual Status Report. American Council on Education, Washington, DC.

Jacobi, M., 1991. Mentoring and undergraduate academic success: a literature review. Rev. Educ. Res. 614, 505–532.

Patton, L., Harper, S., 2003. Mentoring relationships among African American women in graduate and professional schools. New Dir. Stud. Ser. 104, 67–78.

PROFESSOR SUPPORT NETWORKS

12

Professor Support Networks

Pauline Mosley

Seidenberg School of CSIS, Pace University, Pleasantville, NY

We have all heard the phrase *it is not what you know, but who you know*. We all know of someone who is a tenured professor, chair of a department, dean of a school, or president of a university, and we wonder how did she or he get there? We think to ourselves, I am better qualified and more knowledgeable than she or he. The truth of the matter is that successful advancement within the academy depends less on credentials and knowledge and more on an informal web of contacts. Don't get me wrong, one does need a doctorate to teach and know their field to enter into academy ring. But, to advance, to be acknowledged, to be appreciated, to be heard, to be valued—this has a lot to do with who you know and your interactions with who you know. Your connections between people and your ability to foster relationships that cultivate good communication, awareness, trust, and decision making are critical in determining if one survives or thrives.

WHAT IS A PROFESSOR SUPPORT NETWORK?

A Professor Support Network (PSN) is a pictorial diagram which shows people, groups, and organizations and their connections to you and what they can do to support you in achieving your goals in becoming a professor. The nodes in the network are the people and groups while the links show the type of relationship. A solid line between the groups or nodes denotes a very high frequency of interaction. A dashed line between groups signifies low communication or interaction between the nodes. An effective PSN will have all solid lines as oppose to broken ones (-----) signifying weak connectivity. It is better to have a few nodes in a PSN with a strong connection, implying that you are in contact with them at least three times a week as oppose to many nodes with varying connections

171

and just one or two strong connections. The PSN is different than a social network. In a social network, a person may have 50 or more connections or direct ties. The social network is one can interchange information or share ideas and you are expected to contribute "something" to the social network. But, in the PSN these connections are **giving-connections** every node is providing something for you, you are NOT expected to give anything but to accept the support in whatever way it comes. If the node is not giving you something, then it doesn't belong in your support system. Every node in the PSN are nodes that you trust, respect, and will utilize as a sounding board when it comes to making a decision. The PSN has a life cycle just like many other systems. As your goals change, your support network will vary. For example, the support network system for a doctoral student is slightly different than the support network system for an assistant professor aspiring to securing tenure. A robust PSN is a network in which two nodes or more remain the consistent throughout one's transition from student to faculty to chair to dean. The PSN should be reviewed periodically and adjustments to nodes made (Figure 12.1).

In the above figures note that the Ph.D. support network and the tenure and promotion network vary slightly. The family and mentor nodes are consistent as one progress upward and the shift from study groups to research groups shifts as well. The way you interact in a study group while acquiring a Ph.D. and your ability to thrive regardless of the cultural, diversity, or skill-set mix is good preparation for when you transition into the research groups. As students, we all have a tendency to despise working in groups. So-and-so doesn't do their share fair of the work, so-and-so cheats. These individuals will always be there in a group setting. You need to learn how to work with them and not let their bad attitudes rub off on you, but continue to learn how to thrive in a group setting.

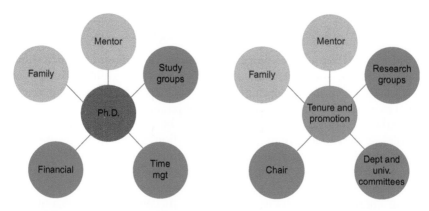

FIGURE 12.1   Ph.D. support network versus tenure and promotion network.

CREATING A PSN

Determining which people, groups, and organizations will comprise your PSN takes time. The first step is to complete the PSN worksheet found in the appendix at the end of this chapter. This worksheet helps you to identify all the people, groups, and organizations that you know and interact with. The next step is to eliminate all people, groups, and organizations that you don't feel will support you or have your best interest at heart. Next, eliminate all people, groups, and organizations that you feel you can't trust. Finally, ask yourself the question what is this person, group, or organization doing for me and how will they help me to achieve my goals?

Research and identify people, groups, and organizations that you wish to be a member of or that you would like to know. There is an old saying *hitch your wagon to a star*. Find at least three people who are where you wish to be and ask them if they will mentor you. Get to know them and utilize them to "pull" you where you wish to go. Developing relationships takes time and patience and developing a PSN will not happen overnight.

Creating an effective and robust PSN should be diverse. Your PSN should be comprised of: males and females; people at your level; people above you; and people way above you. Diversity is critical for it not only provides you with a broad perception but it cultivates a knowledgeable resource base for global or international interactions. Where do you go for information or advice? Where do you go for support and encouragement? Where do you go to vent, cry, or share good news? If it is the same person and only that person and no one else in the PSN, then you might need to rethink your PSN. Another very important consideration is having one global node. This connection to someone in another country or state will bring a different perspective to your goals. In addition, the node will provide you with invaluable insight on issues from a global perspective. The nodes in your PSN should be at least two or three levels if not more above you. If a majority of your nodes are individuals who are on the same academic level as you, this could be problematic. First, they might not possess the insight that you need to advance, second they probably will not encourage you to advance since they haven't done it themselves or don't know how. Finally, you should have at least one node that bridges you to other networks of people aspiring where you wish to go or on an administrative level or higher. Networking is important, and growing your connections is key. Attending conferences, participating on boards and clubs are great. But, you need to constantly make sure that your circle of connections and people in other circles of higher connections are getting to know you and what you do.

FUNCTIONALITY OF SUPPORT NODES

What is support? Support is something that makes someone more determined, hopeful, or confident. It is the word, the encouragement that makes an individual want to go on when the tables are turned and all obstacles are against him. Support can come from family, friends, co-workers, and spiritual avenues. It can come from the young as well as the old and it can come at any time.

Family

What is family support? A family can play an important part in supporting a member who is aspiring for a higher goal in life. There are many family dynamic structures but regardless of the structure or make-up of the family, these individuals closest to you can offer you the best encouragement mentally and emotionally needed for this journey. It some cases, it may be that they have personally attained a certain goal themselves and are reaching down to pull a family member up or the situation could be reverse and they are down and pushing the family member up and above them, most of the times this is the way it is among Latinos and African Americans. How many times have we heard stories coming from the minority communities of "he is the first one to go to college" or "she is the first in the family to get a degree?" These achievements are won because some families have banned themselves together to support the junior members and see that they advance further. In a real family there is no jealousy but instead when the junior member graduates the feeling of accomplishment is shared by one and all.

Dr. Ben Carson and his brother were encouraged and supported by their mother to pursue a higher education. Although she only had a third grade education she encouraged her sons to go further. She could barely read yet she assigned them book reports and thus they became avid readers and began to excel in school. These studies propelled them on to win awards and scholarships and thus become great in their fields.

Donald Thornton was a ditch digger but he had a dream that his six daughters did not have to be low paying nannies but they could be doctors. In the book entitled: The Ditchdigger's Daughters, Donald tells his daughters *I'm not always gonna be around to look after you, and no man's gonna come along and offer to take care of you because you ain't light-skinned. That's why you gotta be able to look after yourselves. And for that you gotta be smart* (Thorton, 2008). It is amazing what encouragement and family can propel one to do. This seems like an impossible feat at that time but the entire family banned together and with hard work and by leaning on one another they realized their dream. Each one became a doctor earning honors and excelling in many fields.

After the birth of my first son, Marcus Paul Mosley who was born prematurely at 26 weeks on December 30, 1994, and feeling sad about missing Christmas, my husband told me that our next door neighbors Fred and Georgeann Krell were keeping up their Christmas decoration until I arrived home from the hospital. My spirits soared and I was just thrilled! I would be having a Christmas after all. True, to their word they kept their Christmas decorations and cheer up until March of 1995 and invited Paul and me over for dinner to celebrate Christmas. This has had a tremendous impact on my life and we fondly call them Aunt Georgeann and Uncle Fred—they are a true integral part of our lives and we have been celebrating Christmas together for over 20 years! Whenever, I hit a roadblock I remember their great act of kindness and it gives me the boost I need to continue. Kindness and encouragement can change a person's life, the Krells' unselfishness have certainly changed mine.

Keep Humble

The way up is the way down. In one's pursuit to higher achievements it is best to climb the ladder of success humbly. To ascend with an inflated ego can turn people off and perhaps help you could receive is withheld. If you begin to think within yourself that your present accomplishments have placed you a step above your family and those who have helped you to reach there, this is bad thinking. If you begin to feel they are not intelligent enough to talk to or you feel too important to be around them, then you are shutting down one of your main support systems.

Different Ways Families Can Give Support

Family support can come in many ways. It can be as simple as a pat on the back, a telephone call, an encouraging card, or a cooked dinner. I have found that when preparing my papers (my dossier or my defense) it has been so helpful to have family members to fill in areas where I found myself unable to cover. Support can be having someone to help pick up the kids from school, help the kids do their homework, help with the household chores. Support can even be a companion noticing that you are on the verge of being overwhelmed and taking the time to whisk you away for an evening night out so that you can emotionally regroup. Support can even come when your little one gives you thumbs up and smile. All these are examples of family support.

Family—Lending an Ear

Family support can be just someone to lend a listening ear. Sometimes when you have been overlooked, disqualified, misjudged, or passed over because of gender, ethnicity, etc. or because you just failed you need a listening ear. You need someone to vent your feelings to. A family member can listen and perhaps help defuse the situation. They can tell you if you are right or wrong. They will tell you that even if you are right you cannot respond with anger and hostility. To get what you need often you must navigate around a situation rather than meeting it head on. The Bible tells us "A soft answer turneth away wrath." Family support will give you help, plans, and strategies to reach your goals. Although both of my parents never received a college degree it certainly did not prevent them from understanding my day-to-day challenges nor giving advice or simply just listening. Both mom and dad would listen to the countless tales of this or that. For me it was a great relief just to articulate what new situation I had found myself in. Together, we would strategize a plan for navigating through the situation. When I married my husband, he would have the honors of listening to me talk for hours on end of various situations and myriads of solutions. On many occasions, I think just to get away from my talking he would tell me to see what my parents thought. Talking with family is a great way to really see after you say what has happened — that things really are not as bad as you think they are. Family, is a great resource for listening and for bringing situations into perspective as well as giving you the support and courage you need to make a decision and move forward. It is amazing how basic principles can apply to even the most complicated situations. Family support may advise you to stand down and let this situation pass over or stand up and speak out for your rights. What would we do without our families?

Family Intervention

A classic case of family intervention happened to my mother-in-law, Deborah Mosley. Deborah was in the last few weeks of her last term of college pressing to receive her Bachelor of Science degree. She was also in her last month of pregnancy. The daily grime of traveling to and from school, the constant effort of trying to keep up with all the assignments of school and house work, plus caring for her husband and small son had worn her down so she came to the conclusion she could no longer go on. "I quit," she said and nothing said could convince her to stick it out. Her husband gathered her together and insisted that they have a private conversation with her mentor and advisor. To Deborah's surprise as they talked the elderly mentor leaned in and looked Deborah in her eyes and said, "My dear, I have been holding on to life just to see you graduate." Deborah was so convicted by her words that she went back to school and 3 weeks later she graduated with her class. It took intervention of both husband and the elderly advisor to support and encourage Deborah to hold on. A week or two later Deborah had a bouncing baby boy.

Having my mother-in-law as a role model of a strong, intelligent, ambitious black woman has helped me tremendously. She was a pioneer in our family who blaze a trail in the STEM field as a nutritionist and dietician, mom knew the ropes and this helped me tremendously in my journey as a mother and as student pursuing education. Mom, knew just what to say and really understood the struggle it is for me to juggle family, work, and school all at the same time. It is a blessing to have someone in your immediate circle who has achieved a level of success and truly understands the struggle, because they not only are able to provide support but they understand the significance of you needing the support.

Co-Workers

Co-workers are important and if you support them they will support you. Instead of competing with your fellow co-workers allow yourself to compliment them in their endeavors on the job. Don't make the job stressful by gossiping about your fellow workers. Don't hog all the credit for a job which the group has done together instead share the glory and even let your follower worker take top billing if they have done more. In time your good attitude may pay off when you need someone to speak a good word for you. This applies to your boss also. Give a full day's work for a full day's pay. Be on time, limit personal calls, and always do your best. Don't steal the boss's time shopping or standing around taking long breaks.

I was fortunate to have an extremely compassionate interim dean, Dr. Connie Knapp and kind fellow workers when I suffered a devastating illness 3 years ago. It was during this time that I had to take a semester off to recover and it was Dr. Knapp along with my chair, Dr. Farkas who encouraged me to use the time to prepare my dossier and apply for promotion to full professor. It was their words of encouragement that made me apply. With my chair's advice and with the graciousness of Dr. Susan Feather-Gannon, who lent me her dossier, along with the support of Dr. Hale (my former chair) I went forward. A sick leave that perhaps would have been spend moping and consumed in my illness was revolutionized. Instead I found myself distracted from my sickness and involved in preparing my dossier to submit for full professorship. My boss and my co-workers stood by me all the way and I give them credit for helping me to obtain my present position. The following semester, I returned as a full professor and my wonderful office mates—Dr. Jean Coppola and Dr. Susan Feather-Gannon welcome me back! It was good to be back. They are the best office mates anyone can have. Both have a wonderful sense of humor, which makes coming to work a pure joy. It is so easy to say, I care—but my colleagues really showed me just how much they care beyond the scope of their jobs—and their acts of kindness I will always treasure for the rest of my life. I thank you all!!

Socializing with Friends

Maintaining a social circle is extremely necessary for it is by networking you make contacts that can help you in your progress. Never pass up an opportunity to socialize, contribute, fellowship with co-workers and friends. It is amazing how interactions in these circles often pay off. Never become so embedded in your studies that you forget your friends. Sidekicks, old pals should not be forgotten. Always find time to schedule them into your calendar. A call, a date to get together is good. Besides making social and academic contacts etc. remember all work or study and no play makes Jack a dull boy. It is a wonderful feeling to have a friend send you a card or give you a phone call just to say hi or give you a word of encouragement. I have been truly bless to have many, many church friends and friends outside of church encourage me along the way. If I were to speak about them.....it would easily add another 100 or so pages more. But, I thank them all for being in my life and being such an encouragement to me.

Spirituality

It is good to be able to draw strength from one's faith and belief in God. Nothing is so positive and uplifting as when mortal can look beyond themselves and receive peace, comfort, and joy from the eternal Creator of all things. It can be very discouraging when you are being honest, working hard, and doing the morally and ethically right thing and you discover that those who you work with are corrupt and yet recognized for their achievements. This can be very puzzling. It is at these crossroads of confusion- where it seems like nothing is making sense and you search for answers or rationale and none exist – that spirituality brings clarity of mind and peace to my soul. I have found when faced with complex problems, difficult decisions, it is wonderful to be able to commit these items to a source greater than me. When I turn

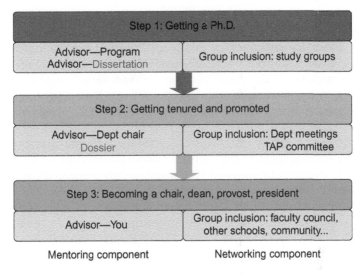

FIGURE 12.2   The process: advancing in the academy.

them over it helps to relieve stress and I find a calm and peace coming over me that truly passes all understanding. During my meditations Scriptures comes to me such as "I can do all things through Christ who strengthens me." Or "there is nothing too hard for God." Words like these encourage me to go on.

THE PSN AND THE ACADEMY

There are many ways to advance within the Academy. Although many universities and institutions have a generalized process and mechanism for advancing, the authors of this book wish to share with the reader a more practical guide and realistic expectations for advancing within the academy. In our opinion, there are three basic steps that must be completed in order to be termed "successful."

Mentoring and Networking

Mentoring (Figure 12.2). In this process we present a schematic depicting three basic steps needed for advancement within the Academy. There are two major components that are critical in completing all of the steps, namely mentoring and networking. One of the most pressing issues facing American universities is the number of students who fail to graduate (Creighton, 2006). Graduation statistics reveal that approximately 26% of students who enroll as freshman do not re-enroll as sophomores (The Chronicle of Higher Education [2005–2006]), and further, approximately 52% of students who entered college actually completed their programs after 5 years (American College Test [ACT], 2002). In spite of all the programs and services to help retain students, according to a government source, only 50% of those who enter higher education actually earn a bachelor's degree (U.S. Department of Commerce, Bureau of the Census Digest of Educational Statistics, 2004).

Though these alarming figures come from undergraduate programs, equally alarming figures surface from within doctoral programs in educational leadership. Smallwood (2004) poignantly indicated the attrition rate in doctoral programs could be as high as 50%. There is some evidence (Lage-Otero, 2005) suggesting women and minorities are leaving their doctoral programs in even higher numbers. Based on their survey of 9,000 students from 21 doctorate-granting universities, Nettles and Millett (2006) indicated a substantive mentoring relationship with a faculty member positively affects progress toward the degree and more importantly significantly relates to completion of the Ph.D. or Ed.D. Further, they reported that 70% of graduating doctoral students have a mentor.

Success in achieving a Ph.D. depends upon a close and effective working relationship with one's advisor and mentor. And yet, while virtually every doctoral student has a research advisor, survey data from the Ph.D. Completion Project and other studies show that not every student has access in their doctoral program to someone they consider a mentor (Nettles and Millett, 2006).

Though mentoring is often cited as among the most influential factors on degree completion, that influence is difficult to assess. Student differences in cultural background and field, or discipline, may result in differing perceptions of effective mentoring. For some students, the mentoring that is valued most may be guidance on dissertation research; for others, it may be advice about how to navigate a career path after completing the degree; and for others, it may mean providing support and counsel when students are experiencing tough times, including such common obstacles as writer's block, complications in the relationship with one's research advisor or committee, or discouraging experiences on an academic job market.

Mentoring is also an area that can pose unique challenges to universities seeking to implement program-level or university-wide improvements. For example, while research supervision is a formal responsibility of graduate faculty, and is recognized as such within the administrative structure and tenure and promotion processes for faculty career advancement, often universities do not have similar formal structures to require and encourage "mentoring," which is sometimes thought of as going above and beyond the call of research supervision duties [2]. Indeed, some faculty may cling to notions that the qualities of good mentoring are either inborn character traits or else habits that are best acquired and instilled informally (e.g., by example) rather than through professional development. Because mentoring is practiced and valued unevenly in doctoral programs, and because student expectations of mentors differ, it is not surprising that students report having unequal access to quality mentors as they pursue their Ph.D. Some students describe their having access to good mentors in terms of "good luck" [3], by contrast to access to their research advisors which is an expectation and understood to be required for degree completion. Some students may have an advisor who effectively doubles as a good mentor, while others may find a faculty member aside from their advisor who can provide additional guidance on research, career, and other topics. Students also report receiving valuable mentoring from their peers in the program as well as from persons outside their department.

Cultivating the skill sets to selecting a "good" mentor is a process that one learns during step 1, only to be repeated in steps 2 and 3. Seeking tenure and promotion or high levels of advancement can be smooth-sailing if one has the "right" mentor. Being adequately informed of one's options and what is expected of one is critical in succeeding. Oftentimes this knowledge can only be learned through a mentor. There have been occasions, where I have met Ph.D. students who attempt to "advise" themselves, because they are so frustrated with trying to find an advisor. But, because Ph.D. program change and there are course stipulations which may not be reflected on the program's brochures or websites this can be not only time-consuming but also disastrous. I have also met junior faculty seeking tenure who failed to seek an advisor within their department apply for tenure and fail, simply because they did not know they needed a certain amount or type of publications, service-related work, or participation in department and university committees. Again, this could have been avoided if they had a "good" mentor. Lastly, seeking higher levels of advancement also merits a "good" mentor, someone who knows the political climate of the college and temperature of the faculty and what areas need to be addressed in order for you to succeed. An experienced and learned mentor can offset your shortcomings and assist you in being successful. Likewise, an inexperienced and unlearned mentor can cause you to fail even when you are adequately prepared and highly qualified.

Networking is the relationship building among people with similar interests and goals. Networking is not a one-time event nor is it a one-sided approach where you never give anything back. Rather, it is a strategy that you can utilize to manage your career process. As your connections increase and you benefit from others, there will be times when you will become a resource for others. Networking is a great resource for getting advice and assistance and for you to give back when needed.

The feeling of isolation and not belonging to a "study group" among doctoral students is a major factor that contributes to the high attrition rate at doctoral programs. Yet despite this recognition, the feeling of isolation has yet to be addressed fully in the design of some doctoral programs. In other words, most programs do not include specific design features that help to handle this feeling among matriculated students (Bess, 1978; Hawlery, 2003; Lovitts and Nelson, 2000).

In a study conducted by Dr. Ali, he noted that the feeling of isolation takes place at different stages in the doctoral program and is manifested in various ways. There are two particular issues that contribute to the development of isolation feeling among doctoral students. First, students begin feeling isolated because of confusion about the program and its requirements. What may start as simple confusion about the program or the requirements of the program quickly grows into a feeling of being left behind and overwhelmed.

Second is the lack of (or insufficient) communication that may take place during various phases of the program. Lack of communication takes place on two fronts: student-to-student and student-to-faculty communication. The basis for isolation revolves around these three issues: lack of communications, miscommunication, and confusion. Furthermore, isolation is felt differently at various stages in the doctoral program. Attaining the doctoral degree involves a different journey than prior to those taken in the pursuit of Bachelors or Masters degrees. Therefore, a different set of intellectual and psychological demands is placed on the students. Hawlery (2003) explain the difference of both demands: in most disciplines, the Ph.D. is considered a research degree and means that its primary purpose is to not prepare practitioners, clinicians, and teachers, but to produce scholars. If you want to be considered a scholar, you must do research.

The adjustment process as noted places a psychological burden that some students may find themselves unprepared for. This combined with the fact that most doctoral departments leave students to deal with the psychological aspects of adjustment to themselves. So the students, who are less prepared for this psychological adjustment, may find themselves left behind, isolated and as a result, may drop out of the program. Given that each student takes the exam alone, in isolation, it separates each individual student from another contributing to feeling behind, overwhelmed, and isolated.

The proposal stage includes selecting a topic of research that will be the focus of the student's research in the dissertation stage. It may also include selecting a faculty advisor, a dissertation committee, and doing a proposal presentation in front of the committee. The student is faced with many topics that he/she has had to narrow down in order to reach a topic that is manageable in the eyes of the dissertation committee. Hawlery (2003) explained the difficulty at this stage: "You are surrounded by ideas, many of which would make an interesting topic. Ideas leap from the printed page, they fall like pearls from the lips of speakers, and a few are even exciting enough to awaken you in the middle of the night. In retrospect, it all seems so simple... Yet the process of carving a topic from among what seems to be an infinite number of possibilities is anything but simple."

The uniqueness of the topic of the proposal makes each student experience different from the others. This kind of work forces each student to work alone without the support that they received during prior studies and during their earlier stages of their doctoral study. It potentially leads to confusion and additional psychological pressure. As noted earlier, most doctoral granting departments leave all the aspect of psychological pressure to the students themselves (Bess, 1978).

This is the last stage of the doctoral program in which successful completion results in attainment of the doctoral degree. The steps required to complete this phase of the program vary considerably among doctoral programs and for doctoral candidates. But the processes by which they complete it are complicated, long and daunting. Lovitts (2001) explained about this:

> These are complex processes with which most students have little familiarity or prior experience. Students who reach this stage know (or discover) that they must conduct research that distinguishes them from their peers. Most feel inadequately prepared to do this type of research and find themselves unprepared for the writing in the style required for a dissertation. (p. 72)

Time constraints become important at this stage as well. At this stage, students work on their own with the occasional advice by their faculty advisor. As time passes by, the pressure increases on the student (Hockey, 1994). Hawlery (2003) noted the importance of the time at this stage and advises students to "guard your time more carefully than your wallet" (p. 109).

This stage is characterized by the students working alone with their advisor in the absence of extensive daily social interaction and communication with their peers or with other faculty. This is complicated by the time constraint. In most cases, there is no specific agenda to follow and there are no marks by which the students measure their progress.

Understanding and learning the importance of groups is critical in step 1 because how you interact with your fellow-peers sets a tone for how you will interact with your fellow-faculty members in the future.

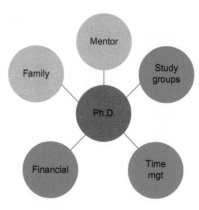

FIGURE 12.3   Ph.D. support network.

Learning how to be social is a vital skill set that definitely can enhance and improve your ability to securing tenure and seeking promotions.

Step 1—Getting a Ph.D.

The first step entails acquiring a Ph.D. or an equivalent doctorate. It is not advised to accept a lecturer position or assistant professorship with the intent of earning a doctorate along the way. This is very hard to do and 90% of the time it is unachievable because although these positions do not require one to conduct research or publish they do require a considerable amount of teaching. Thus, it is very challenging trying to juggle a full-teaching load as well as finding the time to study and complete course work. Lastly, many Ph.D. programs require their students to be full time and will provide a student stipend but do not recommend that students enrolled within a Ph.D. program work full time. I personally would recommend obtaining a Ph.D. first before seeking an entry level faculty position.

There are many factors that affect whether or not one successfully achieves a Ph.D. Figure 12.3 portrays five areas that need to be considered by the Ph.D. candidate.

Pursuing a Ph.D. requires that one has the finances (cash on hand or loan approval), family support, a "good" mentor, a member of a study group, and good time management skills. Failure to have three or less areas supporting you puts you at a very high risk for not completing the Ph.D.

Step 2—Getting Tenured and Promoted

The second step securing tenure demands a plan and a great awareness of one's time table. Every place of higher learning has a designated

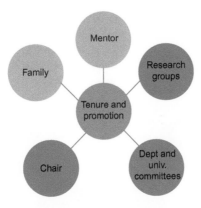

FIGURE 12.4   Getting tenure/promotion support network.

time table in which one must apply for tenure. At most institutions, one must apply for tenure within 4–5 years or request additional time for applying. During this time frame, the faculty is expected to demonstrate proficiency in **teaching** as well as curriculum development, **research** and publish journal articles, attend and present at conferences, and perform **service** to his or her community as well as to their workplace. This requires having an ongoing plan and the ability to multitask such that one meets all of the deliverables in a timely fashion. Documents and letters supporting each of the categories (teaching, service, and research) should be kept on file for inclusion within one's dossier.

Coordinating and preparing the dossier is a major component of becoming tenured. This can be very overwhelming for faculty, especially if they begin preparing it in their last year. This should be updated at the end of each semester to insure that every aspect of your teaching has not been overlooked. You will need the time in your last year to circulate the dossier to the deans of the schools and other selected personnel for their feedback and comments, thereby giving you the opportunity to have the strongest dossier as possible. It is recommended that one utilize the summer to organize, plan, and document the dossier.

Depending on the institution one may be able to apply for tenure and promotion at the same time. Otherwise, they will have to be done separately. Becoming actively involved in one's department as well as networking is essential to having departmental support as well as school support. In the above diagram (Figure 12.4) we depict the importance of learning how to thrive and understand the mechanics of groups. Learning how to become an effective group member can be pivotal to whether one obtains tenure or not. How one behaves and is accepted within a study groups in a Ph.D. program is a precursor to how one may behave within a department. Not only does one need the

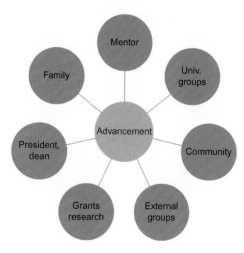

FIGURE 12.5   Career advancement support network.

support of their dean, chair, they also need the support of their fellow colleagues. Seeking out someone within the department who has successfully secured tenure can be most beneficial in understanding the process and what the expectations are for someone seeking tenure. In addition, they may be able to provide you with tips as well as a sample dossier to refer to. Having a dossier that has fulfilled the expectations of the Tenure and Tap Committee is a godsend and I strongly recommend that you inquire within your department about it.

Lastly, knowing who the Tenure and Promotion Committee is comprised of is wise. Making sure that the individuals on this committee know who you are and what your contributions have been can increase your chances of obtaining tenure. It is nice to have a face associated with a name when one is reviewing a dossier. Finally, I can't emphasize the importance of attending meetings. Some people feel that they are boring or don't have the time to attend. Make the time to attend, not only will you learn a lot but more importantly people will get to know you. Attend departmental meetings, faculty council, and join any organization or school-associated activity. This is the fastest way to get known within the unknown, if your schedule permits, be active on these committees. This will serve you well, when you need key people to advocate for your tenure or promotion.

Step 3—Career Advancement

Step 3 involves advancing within the Academy to an administration position. One must be knowledgeable as well a political savvy in order to do this. It is not enough to know your discipline and the school to some extent, you must know people and possess strong communication skills to not only advance but maintain your advancement. I have seen many faculty acquire the position, but then they fail to perform their duties successfully or maintain the position.

Climbing the academic ladder is complex and varied which is why Figure 12.5 has many venues for supporting this effort: family, mentor, universities committees, community, President, Dean, or chair of your department, grants, research, and external groups. Any combination of these supporting groups can propel a faculty to a higher position. Their level of importance is contingent upon your academic culture, your ability to network and articulate what major significant contributions you have made.

A support system can be extremely valuable in ascending the academic ladder. I have found that refusing help or advice from anyone does not mean that you will be a failure—but it will take you along detours that could be avoided. Support is like a ladder, each rung propels one upward. Support is a wonderful commodity that works like money. It takes time to build and grow but eventually when you need to cash it in you will have what it takes to make it happen. Money does not grow on trees and neither does support. Becoming rich doesn't happen overnight and neither does a support system. It is acquired over time with hard work. It is an investment. Invest in yourself and begin working on your support system today, it just may change the course of your life—it certainly changed mine!

References

American College Test, 2002. College Graduation Rates: 1983–2002 Graduation Trends by Institution Type. Retrieved November 2, 2005, Available from: <http://www.act.org/data2002/FileList.html>.

Bess, J.L., 1978. Anticipatory socialization of graduate students. Res. Higher Educ. 8, 289–317.

Creighton, L., 2006. Predicting Graduations Rates at University Council for Educational Administration Public Universities. Sam Houston State University, Huntsville, TX, Unpublished dissertation.

Hawlery, P., 2003. Being Bright is not Enough. Charles C Thomas, Springfield, IL.

Hockey, J., 1994. New territory: problems of adjusting to the first year of a social science PhD. Stud. Higher Educ. 19 (2), 177–190.

Lage-Otero, E., 2005. Doctoral Dissertation: Looking Forward, Looking Backward. Retrieved May 5, 2006, Available from: <http://ctl.stanford.edu/Tomprof/index.shtml>.

Lovitts, B.E., 2001. Leaving the Ivory Tower: The Causes and Consequences of Departure from Doctoral Study. Rowman & Littlefield, Lanham, MD.

Lovitts, B.E., Nelson, C., 2000. The Hidden Crisis in Graduate Education: Attrition from PhD Programs. Retrieved January 27, 2005, Available from: <http://www.aaup.org/publications/academe/2000/00nd/ND00LOVI.HTM>.

Nettles, M., Millett, M., 2006. Three Magic Letters: Getting to Ph.D. John Hopkins University Press, Baltimore, MD.

Smallwood, S., 2004. Doctor dropout. Chron. Higher Educ. 50.

The Chronicle of Higher Education Almanac Issue, 2005–2006. Facts about Higher Education in the United States, each of the 50 States, and District of Columbia, 37–99.

Thorton, S.Y., 2008. The Ditchdigger's Daughters. Dafina Pubishing, New York, NY.

U.S. Department of Commerce, Bureau of the Census, 2004. The Digest of Education Statistics. Thodeore Creighton, Washington, DC.

Index

Note: Page numbers followed by "*f*" and "*t*" refer to figures and tables, respectively.